Lecture Notes in Networks and Systems

Volume 125

The series "Lecture Notes in Networks and Systems" publishes the latest developments in Networks and Systems—quickly, informally and with high quality. Original research reported in proceedings and post-proceedings represents the core of LNNS.

Volumes published in LNNS embrace all aspects and subfields of, as well as new challenges in, Networks and Systems.

The series contains proceedings and edited volumes in systems and networks, spanning the areas of Cyber-Physical Systems, Autonomous Systems, Sensor Networks, Control Systems, Energy Systems, Automotive Systems, Biological Systems, Vehicular Networking and Connected Vehicles, Aerospace Systems, Automation, Manufacturing, Smart Grids, Nonlinear Systems, Power Systems, Robotics, Social Systems, Economic Systems and other. Of particular value to both the contributors and the readership are the short publication timeframe and the world-wide distribution and exposure which enable both a wide and rapid dissemination of research output.

The series covers the theory, applications, and perspectives on the state of the art and future developments relevant to systems and networks, decision making, control, complex processes and related areas, as embedded in the fields of interdisciplinary and applied sciences, engineering, computer science, physics, economics, social, and life sciences, as well as the paradigms and methodologies behind them.

**** Indexing: The books of this series are submitted to ISI Proceedings, SCOPUS, Google Scholar and Springerlink ****

More information about this series at http://www.springer.com/series/15179

Sabu M. Thampi · Elizabeth Sherly ·
Soura Dasgupta · Jaime Lloret Mauri ·
Jemal H. Abawajy · Evgeny Khorov ·
Jimson Mathew
Editors

Applied Soft Computing and Communication Networks

Proceedings of ACN 2019

Springer

Editors
Sabu M. Thampi
School of Computer Science
and Information Technology
Indian Institute of Information
Technology and Management
Trivandrum, Kerala, India

Soura Dasgupta
Department of Electrical and Computer
Engineering
University of Iowa
Iowa, IA, USA

Jemal H. Abawajy
School of Information Technology
Deakin University
Geelong, Melbourne, VIC, Australia

Jimson Mathew
Department of Computer Science
and Engineering
Indian Institute of Technology Patna
Patna, Bihar, India

Elizabeth Sherly
School of Computer Science
and Information Technology
Indian Institute of Information
Technology and Management
Trivandrum, Kerala, India

Jaime Lloret Mauri
Department of Communications
Polytechnic University of Valencia
Valencia, Spain

Evgeny Khorov
Moscow Institute of Physics and Technology
Moscow, Russia

Institute for Information Transmission
Problems (IITP)
Moscow, Russia

ISSN 2367-3370 ISSN 2367-3389 (electronic)
Lecture Notes in Networks and Systems
ISBN 978-981-15-3851-3 ISBN 978-981-15-3852-0 (eBook)
https://doi.org/10.1007/978-981-15-3852-0

This Springer imprint is published by the registered company Springer Nature Singapore Pte Ltd.
The registered company address is: 152 Beach Road, #21-01/04 Gateway East, Singapore 189721, Singapore

Preface

The International Conference on Applied Soft Computing and Communication Networks (ACN'19) organized by Indian Institute of Information Technology and Management-Kerala (IIITM-K), Trivandrum, was held in Trivandrum, India, during December 18–21, 2019. The conference served as a forum for exchange and dissemination of ideas and the latest findings in the aspects of computing and communication networks.

All submissions were evaluated on the basis of their significance, novelty, and technical quality. A double-blind review process was conducted to ensure that the author names and affiliations were unknown to the TPC. This volume contains 16 papers selected for presentation at the symposium.

Special thanks are given to the Conference Committee for the commitment to the conference organization. We would also like to thank all the authors who contributed with their papers to the success of the conference. The conference could not have happened without the commitment of the Local Organizing Committee, who helped in many ways to assemble and run the conference. We express our most sincere thanks to all keynote speakers who shared with us their expertise and knowledge. Finally, we would like to acknowledge Springer for active cooperation and timely production of the proceedings.

Trivandrum, India — Sabu M. Thampi
Iowa, USA — Soura Dasgupta
Valencia, Spain — Jaime Lloret Mauri
Geelong, Melbourne, Australia — Jemal H. Abawajy
Trivandrum, India — Elizabeth Sherly
Moscow, Russia — Evgeny Khorov
Patna, India — Jimson Mathew

Contents

Editors and Contributors

About the Editors

Sabu M. Thampi is a Professor at Indian Institute of Information Technology and Management-Kerala (IIITM-K), Technopark Campus, Trivandrum, Kerala, India. His current research interests include cognitive computing, Internet of Things (IoT), authorship analysis, trust management, biometrics, social networks, nature inspired computing and video surveillance. He has published papers in book chapters, journals, and conference proceedings. He has authored and edited a few books. Sabu has served as Guest Editor for special issues in few journals and program committee member for many international conferences and workshops. He has co-chaired several international workshops and conferences. He has initiated and is also involved in the organization of several annual conferences/symposiums. Sabu is currently serving as Editor for Elsevier Journal of Network and Computer Applications (JNCA), Connection Science - Taylor & Francis, and Associate Editor for IEEE Access and International Journal of Embedded Systems,Inderscience, UK; and reviewer for several reputed international journals. Sabu is a Senior Member of IEEE and ACM.

Elizabeth Sherly received her Ph.D in Computer Science from University of Kerala in 1995, currently working as Senior Professor (HAG) at Indian Institute of Information Technology and Management-Kerala (IIITM-K). She served as the Director of the institute in 2009-2012. Having more than 30 years of experience in teaching and research in Computer Science, her major research work involves Cloud Computing and Machine Learning. She predominantly works in the areas of Medical Image Processing, Natural Language Processing and Automatic Speech Recognition System using Machine Learning and Soft computing techniques. Her Ph.D thesis work was in Artificial Neural Networks in Biological Control Systems now produced Eleven PhDs and actively guiding number of Ph.D Students. She is actively working on promoting virtual learning system with a pedagogy model of Technology Enhanced learning to the Educational Community. Cloud Computing and Marine Wireless Networks and surveillance are her other passionate areas. In Cloud Computing, she works for data replication, migration strategies in SaaS systems. She has been served/serving the Executive Board Member and Technical/Expert member of several Government and Central Government bodies and Editorial Board member of several International and national Journals. Acclaimed with several awards and fellowships which include EURECA (UK) fellowship, Rational Certified (India) Trainer, one among 7 scientists in India for e-Science Grid Group (London) etc. She is the Chief Investigator of two prestigious MeitY projects, Government of India sponsored projects of TDIL (Technology Development in Indian Languages) and one another MICRONet project of ITRA, Media Lab Asia, GOI. Established and heading two centres, Virtual Resource Centre for Language Computing (VRCLC) and Center of Excellence in Pattern and Image Analysis (CEPIA), where Language computing research and Image Processing development using Machine Intelligence are carrying out. More than 100 papers in her credit which includes international, national and conference papers. Her work on Circular Mesh based shape descriptors of MRI breast images for lesion identification had been filed for patent. Editorial member of several journals and conference proceedings, Life Member of CSI, member of IEEE, ACM Chapters, co-authored one book and book chapters also.

Soura Dasgupta is a Professor of Electrical and Computer Engineering at the University of Iowa, U.S.A. He is a past Associate Editor of the IEEE Transactions on Automatic Control, IEEE Control Systems Society Conference Editorial Board, and the IEEE Transactions on Circuits and Systems-II. He is a co-recipient of the Gullimen-Cauer Award for the best paper published in the IEEE Transactions on Circuits and Systems for the calendar years of 1990 and 1991, and a member of the editorial boards of the International Journal of Adaptive Control and Signal Processing, and the EURASIP Journal of Wireless Communications. He received the University Iowa Collegiate Teaching award in 2012. In the same year he was selected by the graduating class for an award on excellence in teaching and commitment to student success. In 1993 he was a selected as a US National Science Foundation Presidential Faculty Fellow, an award given annually to 15 early career engineers and 15 early career scientists for excellence in research and teaching. His present research interests are in Distributed communications, multi-agent control, compressed sensing and stability in aggregate computing. He was elected a Fellow of the IEEE in 1998.

Jaime Lloret Mauri received his B.Sc+M.Sc. in Physics in 1997, his B.Sc.+M.Sc. in electronic Engineering in 2003 and his Ph.D. in telecommunication engineering (Dr. Ing.) in 2006. He is a Cisco Certified Network Professional Instructor. He worked as a network designer and administrator in several enterprises. He is currently Associate Professor in the Polytechnic University of Valencia. He is the Chair of the Integrated Management Coastal Research Institute (IGIC) and he is the head of the "Active and collaborative techniques and use of technologic resources in the education (EITACURTE)" Innovation Group. He is the director of the University Diploma "Redes y Comunicaciones de Ordenadores" and he has been the director of the University Master "Digital Post Production" for the term 2012-2016. He has been Internet Technical Committee chair (IEEE Communications Society and Internet society) for the term 2013-2015. He has authored 22 book chapters and has more than 380 research papers published in national and international conferences, international journals (more than 140 with ISI Thomson JCR). He has been the co-editor of 40 conference proceedings and guest editor of several international books and journals. He iseditor-in-chief of the "Ad Hoc and Sensor Wireless Networks" (with ISI Thomson Impact Factor), the international journal "Networks Protocols and Algorithms", and the International Journal of Multimedia Communications, IARIA Journals Board Chair (8 Journals) and he is (or has been) associate editor of 46 international journals (16 of them with ISI Thomson Impact Factor). He has been involved in more than 400 Program committees of international conferences, and more than 150 organization and steering committees. He leads many national and international projects. He is currently the chair of the Working Group of the Standard IEEE 1907.1. He has been general chair (or co-chair) of 38 International workshops and conferences. He is IEEE Senior and IARIA Fellow.

Jemal H. Abawajy is a full professor at school of Information Technology, Faculty of Science, Engineering and Built Environment, Deakin University, Australia. He was awarded the higher doctoral degree, Doctorate of Science (DSc.) in 2016, by Deakin University for my outstanding research achievements. He is a Senior Member of IEEE Computer Society; IEEE Technical Committee on Scalable Computing (TCSC); IEEE Technical Committee on Dependable Computing and Fault Tolerance and IEEE Communication Society. Jemal's leadership is extensive spanning industrial, academic and professional areas. He has served on the Academic Board, Faculty Board, IEEE Technical Committee on Scalable Computing Performance track coordinator, Research Integrity Advisory Group, Research Committee, Teaching and Learning Committee and Expert of International Standing Grant and external PhD thesis assessor. Prof. Abawajy has delivered more than 50 keynote addresses, invited seminars, and media briefings and has been actively involved in the organization of more than 200 national and international conferences in various capacity including chair, general co-chair, vice-chair, best paper award chair, publication chair, session chair and program committee. He has also served on the editorial-board of numerous international journals and currently serving as associate editor of the International Journal of Big Data Intelligence and International Journal of Parallel, Emergent and Distributed Systems. He has also guest edited many special issues. Prof. Abawajy is actively involved in funded research supervising large number of PhD students, postdoctoral, research assistants and visiting scholar in the area of Cloud Computing, Big Data, Network and System Security,Decision Support System, and E-healthcare. He is the author/co–author of five books, more than 250 papers in conferences, book chapters and journals such as IEEE Transactions on Computers and IEEE Transactions on Fuzzy Systems. He also edited 10 conference volumes.

Evgeny Khorov received his BS and MS degrees with honors from MIPT in 2008 and 2010, respectively, and PhD degree in Telecommunications from IITP RAS in

2012. His PhD thesis focused on Quality ofService (QoS) provisioning in multihop wireless networks. Evgeny Khorov is also a recipient of MoscowGovernment Prize for Young Scientists (2013) for in-depth study of channel access in wireless multi-hop networks. Currently, Evgeny Khorov is the Head of Wireless Networks Lab, IITP RAS, established within Megagrant Project led by Prof. Ian Fuat Akyildiz. Evgeny Khorov has developed numerous mathematical models of networking protocols and designed several algorithms and protocols, which are described in over 80 papers, two of which were evaluated with the Best Paper Awards from IEEE ISWCS (2012) and Elsevier Computer Communications (2018). Evgeny has also received Scopus Award Russia (2018). Evgeny has led many national and international projects sponsored by academia foundations (RSF, RFBR, Ministry of Science and Education, Government) and industry. For breakthrough results of the joint projects, in 2015, 2017 and 2018 Huawei awarded him as the Best Cooperation Project Leader. Evgeny Khorov is also a voting member and contributor of IEEE 802.11 that develops and standardizes Wi-Fi. In 2015-2016, he designed several improvements, which were included in the 802.11ax standard aka high efficiency WLANs. In 2015 he studied technologies for the Internet of Things as a Visiting Research Fellow in King's College London. In 2016 Evgeny Khorov received Russian Government Prize in Science and Technology for Young Scientists for remarkable results of research and development of wireless networks. As a Deputy Head of Department and Associate Professor at MIPT, Evgeny Khorov is responsible for the Telecommunication Program. He supervises BS, MS, and PhD students. Two of his students, Vyacheslav Loginov and Alexey Kureev, have received awards at telecommunication Olympiads. Apart from that, Evgeny Khorov has developed a novel courses on Wireless Networking Protocols. Evgeny Khorov gives keynotes &tutorials and participates in panels at large conferences (incl. IEEE Globecom 2017, IEEE PIMRC 2017, IEEE ICC 2016, ISWCS 2014, NEW2AN 2018, etc.). He is a TPC Chair of IEEE Globecom 2018 Workshop on Cloudified Architectures for 5g abd beyond Systems, IEEE BlackSeaCom 2019, Executive Chair of WiFlex 2013. He serves as an expert of Russian Academy of Sciences and Russian Science Foundation, and also as an editor for Ad Hoc Networks.

Jimson Mathew is currently an Associate Professor and Head of the Department in the Computer Science and Engineering, Indian Institute of Technology Patna, India. He received the Masters in computer engineering from Nanyang Technological University, Singapore and the Ph.D. degree in computer engineering from the University of Bristol, Bristol, U.K. He has held positions with the Centre for Wireless Communications, National University of Singapore, Bell Laboratories Research Lucent Technologies North Ryde, Australia, Royal Institute of Technology KTH, Stockholm, Sweden and Department of Computer Science, University of Bristol, UK. His research interests include fault-tolerant computing, computer arithmetic, Machine Learning, and IoT Systems and cognitive radio systems.

Contributors

Mohamed A. Alasow King Fahd University of Petroleum and Minerals, Dhahran, Saudi Arabia

Sakshi Arora Faculty of Engineering, School of CSE, SMVD University, Katra, Jammu, India

P. Shanthi Bala Pondicherry University, Puducherry, India

Tarannum Bloch Department of Information Technology, Marwadi University, Rajkot, Gujarat, India

M. T. Chitra Indian Institute of Information Technology and Management-Kerala, Trivandrum, India

Alexandre P. da Silva Federal Institute of Education Science, and Technology of Ceará, Aracati, CE, Brazil

R. Divya Department of Computer Science Engineering, Thiagarajar College of Engineering, Madurai, India

El-Sayed M. El-Alfy King Fahd University of Petroleum and Minerals, Dhahran, Saudi Arabia

Maria Ganzha Warsaw University of Technology, Warsaw, Poland;
Systems Research Institute, Polish Academy of Sciences, Warsaw, Poland

B. Gayatri Menon Indian Institute of Information Technology and Management-Kerala, Trivandrum, India

Vineetha Mary Ipe Indian Institute of Information Technology and Management-Kerala, Trivandrum, India

Mohita Jaiswal The LNM Institute of Information Technology, Jaipur, Rajasthan, India

Kamalesh Khanna Department of Computer Science and Engineering, National Institute of Technology Tiruchirappalli, Tiruchirappalli, Tamil Nadu, India

Satish Kumar School of CSE, SMVD University, Katra, Jammu, India

Raja Lavanya Department of Computer Science Engineering, Thiagarajar College of Engineering, Madurai, India

Pascal Lorenz IRIMAS Laboratory, University of Haute Alsace, Colmar, France

S. Mary Saira Bhanu National Institute of Technology Tiruchirappalli, Tiruchirappalli, Tamil Nadu, India

S. Mercy Shalinie Department of Computer Science Engineering, Thiagarajar College of Engineering, Madurai, India

Miloud Mihoubi EEDIS Laboratory, University of Djillali Liabes, Sidi Bel-Abbes, Algeria;
University of Ibn Khaldoun, Tiaret, Algeria;
LabRI-SBA Laboratory, High School of Computer Sciences, Sidi Bel Abbès, Algeria

Harshit Mogalapalli Department of Computer Science and Engineering, National Institute of Technology Tiruchirappalli, Tiruchirappalli, Tamil Nadu, India

Salahadin A. Mohammed King Fahd University of Petroleum and Minerals, Dhahran, Saudi Arabia

Sasmita Moharana Department of Computer Science and Engineering, National Institute of Technology Tiruchirappalli, Tiruchirappalli, Tamil Nadu, India

Mário W. L. Moreira Federal Institute of Education Science, and Technology of Ceará, Aracati, CE, Brazil

B. Nithya Department of Computer Science and Engineering, National Institute of Technology Tiruchirappalli, Tiruchirappalli, Tamil Nadu, India

Tapaswin Padhy International Institute of Information Technology, Bhubaneswar, India

Marcin Paprzycki Warsaw University of Technology, Warsaw, Poland;
Systems Research Institute, Polish Academy of Sciences, Warsaw, Poland

Syam Kumar Pasupuleti Institute for Development and Research in Banking Technology, Hyderabad, India

Silas S. L. Pereira Federal Institute of Education Science, and Technology of Ceará, Aracati, CE, Brazil

Ricardo A. L. Rabêlo Federal University of Piauí, Teresina, PI, Brazil

Abdellatif Rahmoun LabRI-SBA Laboratory, High School of Computer Sciences, Sidi Bel Abbès, Algeria

Damian Rakus Warsaw University of Technology, Warsaw, Poland

Sajad Ahmad Rather Pondicherry University, Puducherry, India

Joel J. P. C. Rodrigues Federal University of Piauí, Teresina, PI, Brazil;
Instituto de Telecomunicações, Lisbon, Portugal;
Center of Excellence in Information Assurance (CoEIA), King Saud University, Riyadh, Saudi Arabia

Kamil Ruta Warsaw University of Technology, Warsaw, Poland

Kashif Saleem Center of Excellence in Information Assurance (CoEIA), King Saud University, Riyadh, Saudi Arabia

K. Selvamani Department of Computer Science Engineering, Thiagarajar College of Engineering, Madurai, India

Abhishek Sharma The LNM Institute of Information Technology, Jaipur, Rajasthan, India

Vaidehi Sharma The LNM Institute of Information Technology, Jaipur, Rajasthan, India

Elizabeth Sherly Indian Institute of Information Technology and Management-Kerala, Trivandrum, India

T. Siron Anita Susan Department of Computer Science Engineering, National Institute of Technology Tiruchirappalli, Tiruchirappalli, Tamil Nadu, India

K. Somasundaram Department of Mathematics, Amrita School of Engineering, Amrita Vishwa Vidyapeetham, Coimbatore, India

Sunanda Faculty of Engineering, School of CSE, SMVD University, Katra, Jammu, India

K. Sundarakantham Department of Computer Science Engineering, Thiagarajar College of Engineering, Madurai, India

Saloni Sureja Department of Computer Engineering, Marwadi Education Foundation, Rajkot, Gujarat, India

R. Thillaikarasi National Institute of Technology Tiruchirappalli, Tiruchirappalli, Tamil Nadu, India

Tony Thomas Indian Institute of Information Technology and Management-Kerala, Trivandrum, India

Raghuvir Tomar The LNM Institute of Information Technology, Jaipur, Rajasthan, India

Fruit Fly Optimization-Based Reliable Routing Algorithm for Wireless Sensor Networks

T. Siron Anita Susan and B. Nithya

Abstract Routing in wireless sensor network is a predominant task to transmit the packets efficiently from a source to destination. The sensor nodes are scattered in the network and should be self-manageable to route their packets. Due to the stringent resource constraints, the routing needs to be more reliable with minimum energy to prolong the network life time. Considering these points, this paper proposes a novel algorithm, namely reliable routing algorithm (RRA) to find a reliable optimal path from source to destination. To achieve this goal, the proposed RRA algorithm adopts fruit fly optimization approach along with the runtime dynamic decision metrics. The decision metrics such as expected transmission count (ETX), neighborhood status and residual energy are utilized to make the judgment about the reliability of the path. Due to these dynamics, the proposed RRA algorithm effectively controls node failure and link failure, thereby ensuring stable and reliable routing in WSN. The performance of the proposed RRA algorithm is validated with the predicted remaining delivery (PRD)-based routing algorithm and multi-objective fractional particle lion algorithm (MOFPL) in terms of packet delivery ratio (PDR), delay and energy balance factor (EBF).

Keywords Wireless sensor network · Routing · Fruit fly optimization · Reliability · Node failure · Network life time

1 Introduction

Wireless sensor networks (WSNs) have various captivating applications of sensing and monitoring remote locations where human access is not possible. This sensed data should be routed from source to the desired destination by taking care of various

T. Siron Anita Susan (✉) · B. Nithya
Department of CSE, National Institute of Technology Tiruchirappalli, Tiruchirappalli, India
e-mail: anitasusan.t@gmail.com

B. Nithya
e-mail: nithya@nitt.edu

S. M. Thampi et al. (eds.), *Applied Soft Computing and Communication Networks*, Lecture Notes in Networks and Systems 125,
https://doi.org/10.1007/978-981-15-3852-0_1

network and sensor constraints. This makes routing in WSN as a quiet challenging task that leads to big evolving research area. The process of routing protocol is to select the suitable routing path to take away the data from one node to another with the aid of routing table. Due to the characteristics of WSN, the routing algorithm needs to find the routing path as a reliable, stable and optimal path till the destination node. There are different routing schemes based on the number and type of destination nodes. They are: (i) unicast: routing a data from a source node to one desired destination node, (ii) broadcast: routing a data to a set of nodes of a network, (iii) multicast: routing to group of nodes which are interested to receive the data, (iv) anycast: routing from a source to any one of its nearest node among a group of nodes and (v) geocast: routing based on geographic nodes. These categories of routing are depicted in Fig. 1.

It is inferred from the figure that the routing algorithm designed for a particular routing category will not be suitable for remaining categories. This paper aims to give efficient routing algorithm for unicast transmission. Related to this, lot of routing protocols for WSN are proposed in the literature which are broadly classified based on their functioning mode, participation style and network structure [1, 2] as shown in Fig. 2.

With respect to functioning mode, the routing protocols are classified as proactive, reactive and hybrid routing protocols. Proactive routing protocol stores all the routes between a source to its destination in the routing table in prior and transmits the data accordingly. Whereas in reactive routing protocol, the node obtains its desired path only when it needs to transmit the data. Hybrid routing protocols acquire the marks of both proactive and reactive protocols. The node exchanges data among other devices using proactive protocol, if they are within the same region (zone). For outside zone communication, the reactive routing protocol is utilized.

Based on the participation style, the routing protocols are classified as direct communication, flat, clustering. In direct communication, the routing of data happened between the node and base station (BS) without any intermediate nodes. In flat routing, the same functionalities are imposed on each node. It means that all of the nodes

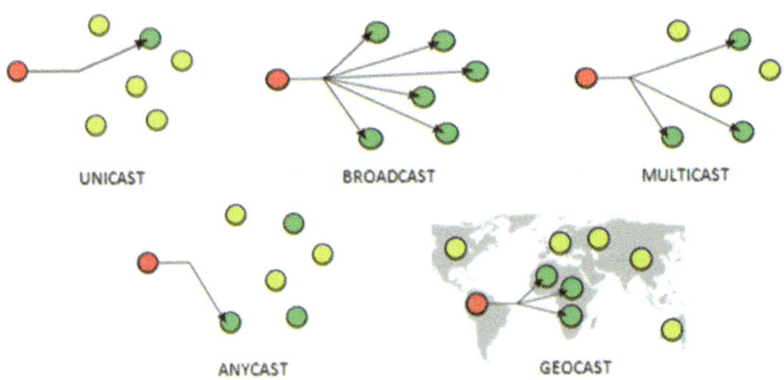

Fig. 1 Different routing schemes

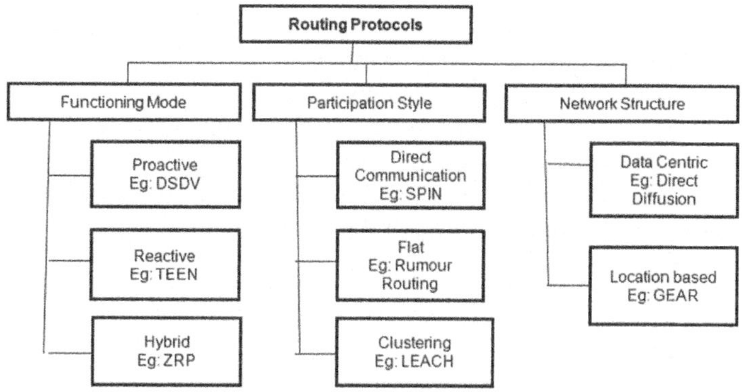

Fig. 2 Classification of routing protocols

in the network are expected to play the same role for the data transmission. In clustering type of routing protocol, the network is divided into clusters with cluster head (CH), and members within the cluster communicate with others through CH.

In terms of network structure, the routing protocols are classified as data centric and location-based routing protocols. In data centric, query-based routing model is adopted. The requesting node sends its query to all other nodes, and nodes which have the requested data will reply back. In location-based routing, location information which is obtained using GPS is utilized for routing among nodes.

There are lot of metaheuristic algorithm to solve a problem optimally, and those algorithms are also used to find an optimal route or path in WSN. The optimization algorithm finds the best solution based on the objective function of the desired problem, and it guarantees to achieve the better solution. The most widely used approaches are ant colony optimization algorithm (ACO) [3], particle swarm optimization (PSO) [4], lion optimization algorithm (LOA) [5], firefly optimization, fruit fly optimization (FFO) [6] and so on. Among these approaches, FFO technique has certain advantages over other surveyed optimization techniques [6]. FFO is fast in solving the problems and is simple structured. It is a stable algorithm with few parameters and its adaptability to the real-time applications. Due to these reasons, the proposed RRA algorithm adopts FFO approach to find the optimal path. Fruit fly is more admirable than other flies, and they can reach the food source by using their smell and vision organs. Initially, the fruit flies use their smell character to find all scents of its food and then fly toward the food. This FFO is developed based on the foraging behavior of fruit fly. The two main phases of this optimization algorithm are smell phase and vision phase. In smell phase, the flies fly closer to their food by using their smell capability, and in vision phase, the flies will go nearer to their food. These two steps are repeated till the fly catches its food. In case of routing, the smell phase identifies the fitness value for an optimal path which is constructed in the vision phase.

Based on the above discussion, it is inferred that routing of the data from a source to its destination needs to be handled in an efficient way by using suitable algorithms to reduce the complexity and delay, thereby leading to more reliability. There are numerous routing schemes proposed in the literature to attain these goals, and some of them are discussed in the next section. The rest of the paper is structured as follows. In Sect. 2, related work is briefly explained. Section 3 gives the proposed work, and finally Sect. 4, provides the conclusion of the paper.

2 Related Work

Due to various changes in the wireless environment, the quality of link between any two nodes plays an important role while routing the data. To enhance the routing and overall network performance, link and node failures need to be minimized. Many of the existing routing protocol uses minimum hop count, distance, link weight, etc., to find the least cost and efficient route. But, there might be some significant lossy link. To overcome this and to predict a high throughput link or path, expected transmission count (ETX) metric is used. Many researchers [7, 8] have used ETX for estimating the link quality between the nodes. ETX is the predicted number of transmission and retransmission of data packet over the link. This ETX is calculated with the help of forward and reverse delivery ratio of a node. The number of packets that the link is able to transmit is said to be forward ratio (FR) of a link, whereas the number of packets sent back as retransmission on the same link is termed as backward ratio (BR) of that link. The ETX of a link is calculated as

$$\mathrm{ETX} = \frac{1}{\mathrm{BR} * \mathrm{FR}} \tag{1}$$

Xiaohan et al. [7] have used a routing metric called predicted remaining deliveries to attain energy efficient and link reliable routing. The weights and delay are assigned to each link, and accordingly, the node status is estimated. Initially, a hello message is broadcasted by each sensor node to its neighbor and is also used to calculate the ETX between sensor nodes. Upon receiving the hello message, each node sends its ID, energy level, ETX value and the delay to its neighbors and is stored in the neighbor table and is updated frequently. Based on the stored values, the score of individual link and entire route is calculated to estimate the route score of each route. Finally, best route score is chosen by the parent node to forward its data.

Singh et al. [4] have proposed a method for routing in a cluster-based WSN where the characteristics of PSO and VLEACH protocol are combined. The basic LEACH protocol is a cluster-based protocol where the cluster head is selected periodically based on the energy level. VLEACH is an advanced LEACH protocol where each cluster will have a vice cluster head. It means that it acts as an alternative cluster head. The data will be routed to the base station when a CH dies. Here, PSO technique is used to optimally choose the CH which has the maximum energy.

ACO [3] is one of the probabilistic approaches in that the ant colony and the behavioral actions of ants toward the search of the food are modeled. Based on this approach, Sun et al. [9] have proposed an efficient routing protocol. Here, transmission distance, direction and residual have been used to choose an optimal routing path. The ants of this ACO select the path according to the pheromone and heuristic value obtained from the gathered information.

LOA [5] is a new population-based algorithm which is derived from the lifestyle and cooperative activity of lions. This algorithm is designed by modeling the characteristics of lion. The initial population is divided as nomad lions and prides. Some percentages of prides are considered as females and rest as males, and this consideration is vice verse in nomad lions. For each member, the best position is obtained by comparing with previously obtained visited position. In this algorithm, an area called pride territory is taken with the best visited position of each member. Also, some female lions are selected randomly for hunting which move toward their food and catch it. Bhardwaj et al. [10] have proposed an algorithm based on particle lion algorithm called multi-objective fractional particle lion (MOFPL) algorithm. In this, an optimal cluster head is chosen and then an optimal routing path is calculated based on multi-objective function. This algorithm combines fractional theory and PSO along with LOA to find optimal path. Based on the calculated multi-objective fitness function, the fitness of each node is calculated in MOFPL algorithm and an optimal path is calculated.

FFO [6, 11] is a new optimization algorithm which gives optimal results based on the food searching characteristics of the fruit flies. Hazim et al. [12] used a modified FFO to find an optimal solution for the well-known traveling salesman problem (TSP). In this problem, the smell values are calculated for the initial arrangement of city, and the best value is stored that will be used in vision phase. The author modified the vision phase by constructing and updating a city arrangement matrix based on the best solution in each generation.

2.1 Overall Inference and Motivation

Based on the above discussion on various optimization techniques used in routing, it is inferred that the FFO has certain advantages over other surveyed optimization techniques [7]. FFO algorithm is simple structure and stable algorithm and is best suited to find reliable routing path. It is fast in solving the problems with few parameters, and it is easily adaptable for real-time problem-solving techniques. Moreover, the reliability of link and the path should also be considered for choosing a reliable link for routing. With this motto, this paper proposes an adaptive algorithm, namely reliable routing algorithm (RRA) based on fruit fly optimization algorithm. The main objective of the proposed RRA algorithm is to combine the reliability in optimal path calculation since the existing algorithm only concentrated on optimal path and not on reliability. To find a reliable and an optimal path for routing, RRA utilized ETX metric for estimating the reliability of the link and uses fruit fly Optimization algorithm to maintain the optimality of the path. The transmission range and energy level of the node influence the proposed RRA algorithm to elect the potential intermediary nodes to route the data.

3 Proposed Reliable Routing Algorithm (RRA)

The main objective of the proposed RRA algorithm is to find an optimal path from a source to the corresponding destination. An optimal path toward the destination needs to have lesser number of connected intermediate nodes with sufficient energy resources. The path is optimally chosen by inheriting fruit fly optimization algorithm where it uses ETX value for each link. The proposed RRA algorithm works in two phases, namely refining phase and construction phase. In the refining phase, the capable and potential nodes are elected. This means that all the elected nodes are within the transmission range of source and have the energy level greater than the threshold value. This threshold energy value is the minimum required amount of energy needed for data transmission. The construction phase is further subdivided as smell phase and vision phase. The smell phase is completed with the determination of ETX for the links of the chosen nodes. With this vital information, the vision phase is started to find out an optimal path using fruit fly optimization approach. The complete overview of the fruit fly optimization algorithm is depicted below:

1. In every optimization problem, the number of generations and the convergence of result at every generation are required to be determined.
2. Every generation will have group of members called as population.
3. Each member will be the possible solution for the fruit flies. In the proposed algorithm, every member is considered as a path. It may not have all the nodes and may not be of same length.
4. For every member, the fitness value (smell function) is calculated and from that a path with the highest fitness value is selected.
5. In vision phase, the fitness value is sent to every member which is obtained from smell function.
6. Each member will be interpolated with itself based on the rules proposed by optimization algorithm. Using that, a new member for next generation is generated.
7. After all members are sent to second step, the newly generated members (population) need to be sent to the next generation, and the entire process is repeated.
8. This process is terminated when a path is obtained with the highest fitness value for more number of consecutive generations and declare that path is an optimal path toward the destination.

3.1 Phase 1: Refining Phase

In this phase, the initial refinement of the network is performed. The network N_i is refined based on the transmission range and remaining energy level, and the resultant network N_s is then used to carry out for further processing.

Initially, hello or probe packets are sent continuously to all one-hop neighbors, and based on the number of transmitted packets, the FR is determined. Once the packet is received by the destination, it should acknowledge the received packet along with its remaining energy level. With the aid of the number of acknowledgment, the BR of the link is calculated. The nodes that receive the acknowledgment check for their energy level of the neighbor nodes which are within its transmission range with a minimum threshold value. The node will choose all its neighboring nodes that have sufficient remaining energy for further transmissions. The energy-deficient nodes are neglected as depicted in Fig. 3. Initially, among 7 nodes of a network N_i, nodes 2 and 6 are neglected due to lesser energy level which is less than the threshold value Th. The remaining nodes of the network N_s are chosen, and the links of the neglected nodes are also disconnected. These nodes of the N_s are renamed sequentially. Based on the estimated forward and backward ratio, the ETX value for the links is determined. This ETX value will give the metric score value for a good reliable link with which a reliable link is calculated as mentioned in Eq. (1).

3.2 Phase 2: Constructing Phase

The construction phase is further subdivided as smell and vision phase. The smell function checks the calculated ETX value and considers the highest reliable value in its state. This chosen value is the fitness value of this smell function. Each generation calculates its fitness value. The vision function checks for the maximum optimal fitness value and gets updated toward the best reliable solution of that generation.

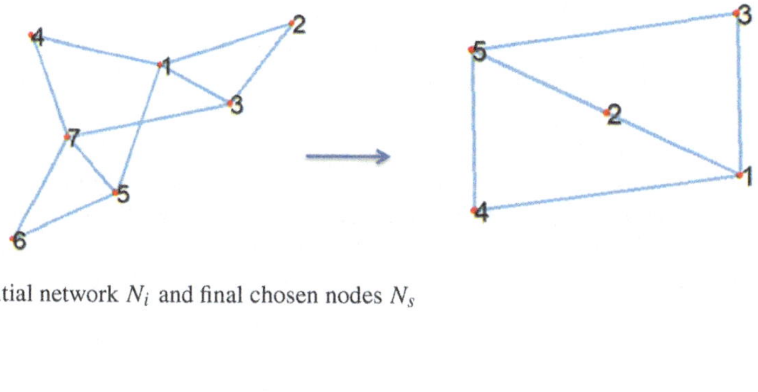

Fig. 3 Initial network N_i and final chosen nodes N_s

The initial population(pt) is generated with all possible paths among n nodes. In each generation, the best ETX value is chosen, and based on that value, the path is calculated in vision phase. To get an optimal path, this process is repeated till the ETX value is converged to the same value. Once, it is congregated, the obtained path is considered as optimal path which has sufficient remaining energy to route the data. Let the variables ETXT, ETXG and ETXMAX denote the ETX value of the path (members), the maximum ETX value of a member in present generation and the ETX value of member with respect to all generation, respectively.

Sub-phase 1: Smell phase- For the valid paths in the population(pt), the ETXT is calculated as given in the proposed Eq. (2).

$$\text{ETXT} = \text{ETXT} + \text{ETX}[\text{path}[i] - 1][\text{path}[i+1] - 1] \tag{2}$$

where ETXT calculates the ETX value of the nodes which are getting added each time in the optimal path. path$[i - 1][[i + 1] - 1]$ represents the link from one node to another for which the corresponding ETX value to be added to the ETXT. The maximum value of the present generation is calculated, and the corresponding ETXT value is stored in ETXG. The path which has maximum ETX is selected and assigned to ETXMAX. Also, the path corresponding to the ETXMAX is assigned to pmax.

Sub-phase 2: Vision phase- For all paths in the population, 30% [11, 12] of the nodes are selected in the path and checked if it is connected to the rest of the nodes in the path. If the selected nodes are found to be connected to the rest of the nodes, then arrange the selected node consecutively to the connected node. The path result is recorded in temp. If there is any node to be disconnected, the corresponding node is deleted and the resultant value of temp is stored as the new value in population at the same position. The first 30% (except the source node) of the best P path is chosen, and the nodes sequence of this solution is applied to the neighbor solutions in connected manner. On considering the path for reinsertion, taking the ceil of the 30% cut value and is given as

$$\text{Cut} = \text{ceil}(\text{length of path} * 0.30) \tag{3}$$

Consider a node Y from 30% and compare with existing nodes in path, one by one. If a node X is connected to the node Y (which we selected from 30%), then insert node Y to next of the node X, else compare with next node in the path. Repeat this step for all nodes in 30%. If the node Y is not connected to any one of the nodes in the path, then neglect that node. Make the resultant path as P1. Now consider path P1; check X_i node is connected to $X_i + 1$ node, if it is connected, continue else remove the node $X_i + 1$ from the path. Repeat this step to all nodes in the path. Finally, the optimal path is constructed. This path results in the best ETX value comparatively to all other paths. The complete code of the proposed RRA algorithm is given in Algorithm 1.

Algorithm 1: Reliable Routing Algorithm (RRA)

Input: ETX matrix(M_i), Initial population(pt), Transmission range(T_i) and Threshold energy level(Th)

Output: Selected nodes(Ns), Path, Total ETX(ETXMAX), Total generation(gen).

```
Initialization of Network Ni, ETXMAX=0, ETXG, gen=cnt=0 and array C.
Broadcast Hello/probe packets to one hop neighbors.
Calculate ETX as of Eq. (1)
for each N_i do
    if Transmission(N_i)<T_i and energy(N_i)>Th then
        Store in NS
    end
end
Generate pt and initialize M_i for N_s.
repeat
    Call procedure Smell ()
    Call procedure Vision ()
until cnt<15
Procedure Smell ()
for each path pt_i in p do
    Calculate ETX value of connected path of Eq. (2).
    Find maximum ETX value for gen_i.
    Store the calculated ETX value of the path in ETXG.
    Store the corresponding path of the ETXG to optimal path.
end
if ETXG>ETXMAX then
    ETXMAX=ETXG and pmax=optimal path
else
    cnt++
end
if ETXMAX is constant for consecutive 15 generation then
    Display it as the final optimal path.
end
End Procedure
Procedure Vision ()
for each path pt_i in p do
    Calculate cut as of Eq. (3).
    Store all nodes between 0 to cut of path in C.
    for each node N_i in C do
        if N_i is connected to node in path then
            Insert node N_i to new path at j+1th location.
        else
            neglect the node.
        end
        Store new path.
    end
    Neglect the nodes which are not connected in new path.
end
End Procedure
```

4 Simulation and Performance Analysis

The simulation is conducted by MATLAB R2016b on an Intel(R) core(TM) i7-6700 CPU with 3.40 GHz and 16 GB RAM. Initial population is constructed based on all possible paths of N nodes. The number of nodes varies from 5, 10 …20. Since the energy levels of the nodes are calculated in advance, it is assumed that no node will get disconnected due to energy loss and time taken for the death of a node gets increased.

4.1 ETX Convergence Versus ETX Value

Based on the continuous stabilized value of 15 generations, the best ETX and the corresponding optimal path are chosen. If the node density is less than the estimated ETX is converged within the lesser number of generations. However, the lesser number of generations is not feasible as it may lead to same path for several nodes. Due to this, the number of generations is assumed to be as 15 that may be changed depending upon the population. Since the elected path is constant for more number of generations, it will be more stable and optimal. Figure 4 depicts the convergence graph to show that the ETX value of certain path remains constant for 15 generations. The corresponding path is taken as on optimal path, and its corresponding ETX value is an optimal ETX value.

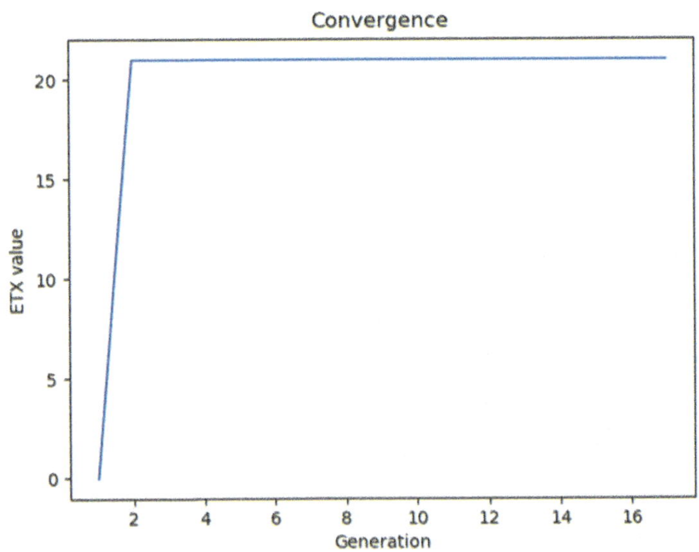

Fig. 4 Optimal ETX value

4.2 Packer Delivery Ratio (PDR)

Packet delivery ratio is the ratio of the backward ratio (BR) of a node to the forward ratio (FR) of that node. It is calculated as follows:

$$\text{PDR} = \frac{\text{BR}}{\text{FR}} \tag{4}$$

Figure 5 depicts the PDR achieved by the proposed RRA algorithm and PRD. Initially, the delivery ratio is 100% and starts to decrease due to the usage of some poor quality links at some situation. If the lifetime of the node is expired due to energy depletion, other sensor nodes which have poor links may get a chance in routing thus leading to drastic decrease in packet delivery. Since the proposed RRA has chosen the average energy level nodes for sending and receiving packets, the lifetime of the node is increased, thereby giving more reliable and stable path. Due to this capability, the proposed RRA attains better PDR than PRD algorithm.

4.3 Network Delay

It is one of the performance metrics which designates the time taken to transfer the data from one node to its destination. It depends on EXT value, transmission and propagation delay, and the delay is calculated as

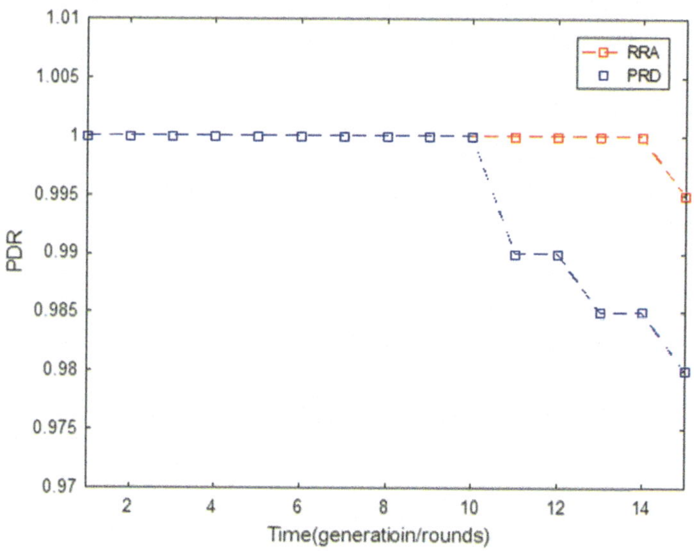

Fig. 5 Comparison of PDR

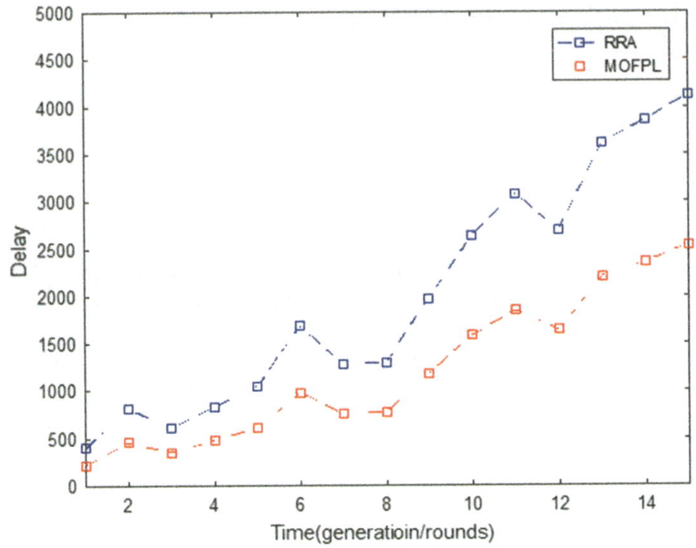

Fig. 6 Comparison of delay

$$\text{Delay} = \sum_{i=0}^{N} Ei(\texttt{Trans\_Delay} + \texttt{Propa\_Delay}(i)) \tag{5}$$

where Ei represents the ETX value of ith node, `Trans_Delay` indicates the transmission delay of the network, and `Propa_Delay`(i) represents the propagation delay of ith node. The transmission delay of the network depends on the packet size and transmission speed. When comparing to Bhardwaj et al. [10] transmission speed of the network, the transmission speed of the proposed RRA algorithm gets increased which results in lesser delay. The transmission speed directly depends on an efficient link performance. The proposed RRA calculates the reliable link for each transmission which increases the transmission speed and results in lesser delay as depicted in Fig. 6.

4.4 Energy Balance Factor (EBF)

EBF is calculated based on the standard deviation of the residual energy of the nodes and is given as

$$\text{EBF} = \sqrt{\frac{1}{N}\sum_{i=1}^{N}(\text{Energy}_{\text{res}}(i) - \text{Energy}_{\text{Avg}})^2} \tag{6}$$

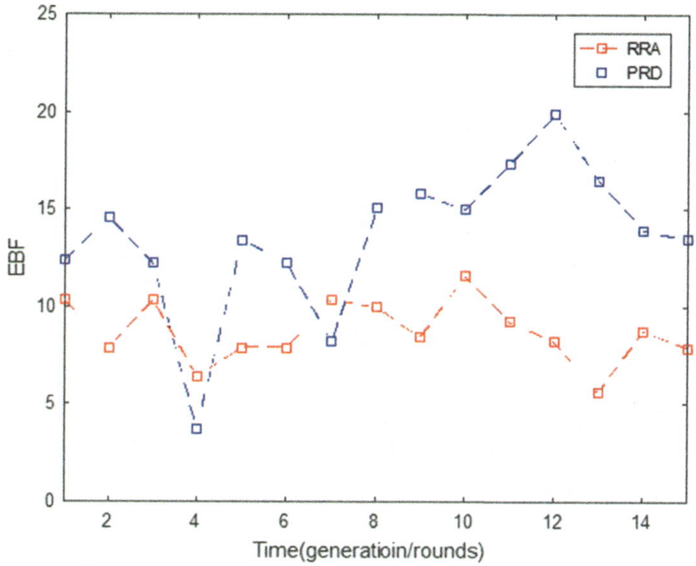

Fig. 7 Comparison of EBF

where N represents the number of nodes, $\text{Energy}_{\text{res}}$ represents the residual energy of a node, $\text{Energy}_{\text{Avg}}$ is the average residual energy of the nodes. The lower the EBF gives better balance in energy consumption of nodes. The proposed RRA gives lower EBF than the existing PRD. The EBF comparison is depicted in Fig. 7.

5 Conclusion

This paper proposes a reliable and an enhanced RRA based on ETX value and FFO to calculate an optimal path in a wireless sensor network. The main goal of the proposed algorithm is to calculate a reliable optimal path. The proposed RRA algorithm yields more reliable and stable path as it utilizes runtime dynamic parameters and inherits FFO approach to make the routing decisions. The results indicate that the proposed RRA algorithm guarantees enhanced PDR with lesser delay. This results in more reliable and stable routing paths.

References

1. Anil P, Jagadeeswararao E, Nagaraju G (2014) A brief study on routing protocols in wireless sensor networks. Int J Res Comput Appl Robot 10–14 (2014)
2. Bhattacharyya D, Kim T, Pal S (2010) A comparative study of wireless sensor networks and their routing protocols. Rev Artic MDPI Sens 10:10506–10523

3. Chandra Mohan B, Baskaran R (2012) A survey: ant colony optimization based recent research and implementation on several engineering domain. Elsevier Expert Syst 39:4618–4627
4. Singh A, Rathkanthiwar S, Kakde S (2016) Energy efficient routing of WSN using particle swarm optimization and V-leach protocol. IEEE Xplore Digit Libr. https://doi.org/10.1109/ICCSP.2016.7754544
5. Yazdani, M., Jolai, F. (2016) Lion optimization algorithm (LOA): anature-inspired metaheuristic algorithm. J Comput Des Eng 3:24–36
6. Iscan H, Gunduz M (2016) A survey on fruit fly optimization algorithm. IEEE Xplore Digit Libr https://doi.org/10.1109/SITIS.2015.55.
7. Lai X, Ji X, Zhou X, Chen L (2018) Energy efficient link-delay aware routing in wireless sensor networks. IEEE Sens J 18:837–838 (2018)
8. De Couto DS, Aguayo D, Bicket J, Morris R (2005) A high-throughput path metric for multi-hop wireless routing. Wirel Netw 11. https://doi.org/10.1007/s11276-005-1766-z
9. Sun Y, Dong W, Chen Y (2016) An improved routing algorithm based on ant colony optimization in wireless sensor networks. IEEE Commun Lett 21:1317–1320
10. Bhardwaj R, Kumar D (2019) MOFPL: multi-objective fractional particle lion algorithm for the energy aware routing in the WSN. Elsevier Pervasive Mob Comput 58:01–16
11. Choubey NS (2014) Fruit fly optimization algorithm for travelling salesperson problem. Int J Comput Appl 107:22–27
12. Iscana H, Gunduza M (2017) An application of fruit fly optimization algorithm for traveling salesman problem. Procedia Comput Sci 111:58–63

User Authentication of IoT Devices for Decentralized Architecture Using Blockchain

Raja Lavanya, K. Sundarakantham, S. Mercy Shalinie, R. Divya, and K. Selvamani

Abstract These days, IoT devices are deployed at an ever-increasing rate worldwide. It is estimated that the growth of IoT devices will increase up to 31 billion devices by the end of next year. Due to this ever alarming increase in cases of identity theft and data leaks all around the world, authentication has become a major concern. As the IoT devices are resource constraint and incapable of defending, they offer potential points of failure within an IoT-based system themselves. Hence, device authentication and authorization is critical for securing IoT systems, especially when the IoT device holds crucial information like medical data or defense-related data. In this paper, authentication for IoT devices using Ethereum smart contract which is a decentralized approach is discussed. Decentralized application instead of centralized system architecture is preferred much. This system makes use of Ethereum smart contracts to ensure secure authentication of device by the respective user. The key aspects of how the system works, its architecture and the implementation of smart contracts on a local blockchain are discussed.

Keywords Ethereum · Authentication · Smart contract · Blockchain · IoT security

1 Introduction

Internet of Things is the developing technology of making all things connected to the Web. These devices can be connected to the Internet and can be controlled and managed through the Internet. The concept of IoT has faced criticism, in regard to the security and the data privacy of the devices which are being used by end

R. Lavanya (✉) · K. Sundarakantham · S. Mercy Shalinie · R. Divya · K. Selvamani
Department of Computer Science Engineering, Thiagarajar College of Engineering, Madurai, India
e-mail: rlit@tce.edu

K. Sundarakantham
e-mail: kskcse@tce.edu

S. Mercy Shalinie
e-mail: shalinie@tce.edu

S. M. Thampi et al. (eds.), *Applied Soft Computing and Communication Networks*, Lecture Notes in Networks and Systems 125,
https://doi.org/10.1007/978-981-15-3852-0_2

customers to cater their needs. The IoT's major significant trend in recent years is the massive growth of devices connected and controlled by the Internet. The wide range of applications for IoT technology means that there is a wide range of aspects in which the IoT devices operate but share some common basics in terms of characteristics and the architecture. The IoT creates immersive opportunities for more and more direct integration of the physical world of devices into digital systems, resulting in effective efficiency improvements, economic benefits such as reducing manual labor and reduced human exertions and automation of the entire scenario in which its being implemented. A large number of IoT devices are utilized at a large scale which leads to a prediction that there might be billions of IoT devices by the year 2020.

Since IoT devices are resource constrained and complexity of devices is less, they can be easily attacked. Thus, it is essential to ensure security of resource-constrained IoT devices by providing proper authentication and access control algorithms [1]. Since they are resource constrained, computation has to be reduced. Also, the authentication scheme must be secure against known attacks and threats. Many IoT solutions have now been put forth by researchers in utilizing the cloud computation power to relieve IoT devices from some of the heavy processing computational workload.

A blockchain is a time-stamped series of immutable records of data which is managed by cluster of computers not owned by any single person. Blockchain gives the Internet users the ability to create value and authenticates digital information in all types of formats. This information can be used in various forms like crypto-currency, smart contracts, crowdfunding, supply chain. The application of blockchain is endless, and also, it is the most secure form available on the Internet to protect the data. Thus, blockchain technology provides solution to most of the IoT security problems.

To date, most authentication techniques for IoT systems are centralized in design and deployment and rely on a trusted third party to authenticate the entities such as open authorization (OAuth) protocol. This approach has drawbacks such as high cost, being a single point of failure, hacking and privacy evasion. To overcome the drawbacks of the centralized authentication, a decentralized authentication scheme using blockchain technology is proposed in this paper which will provide security of IoT devices without the interruption of trusted third party. Additionally, the proposed work involves the use of new architectures including edge computing and blockchain, which facilitates performing authentication at scale. The primary goal of this paper is to propose and analyze the security of a blockchain-based authentication scheme and communication at scale for IoT devices. Especially, we present an architecture and design of a system involving end users, IoT devices, as well as Ethereum blockchain smart contracts which govern the authentication rules and logic and also deliver the data to the device in a secure manner. The primary contributions of this paper can be summarized as follows:

- Proposed a decentralized and scalable authentication mechanism that utilizes blockchain with connectivity to Ethereum smart contracts for authenticating user access to IoT devices, whereby access tokens are issued by the smart contracts with no intermediary or trusted third party.

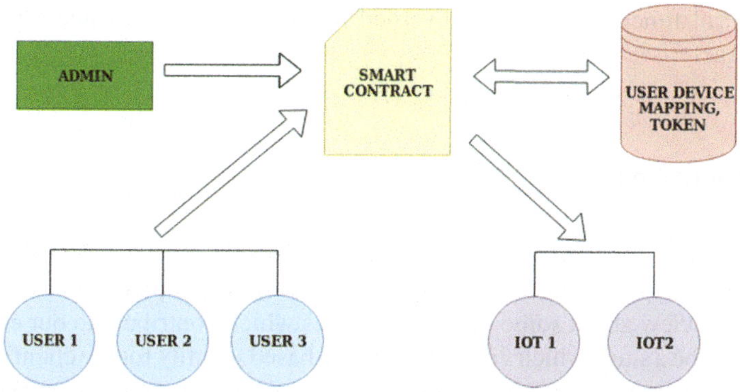

Fig. 1 Architecture of the system

- Presented the details of the overall system including the architectural design, sequence diagrams and interactions among participants with the smart contract including on-chain and off-chain activities.
- Provided the full code of the Ethereum smart contract including the implementation and the main testing scenarios of the overall system functionalities.
- Presented security analysis of our proposed authentication scheme and discussed how the scheme achieves security goals (of confidential, integrity and availability) and is able to overcome eavesdropping, replay and denial of service.

A user authentication system will contain the components that will manage:

- Contract for each end user.
- Contract for each IoT device.
- Ethereum smart contract at a particular address to validate the tokens and deliver the payload of data to the respective device contract (Fig. 1).

This is the proposed architecture of the system to be constructed [2].

For this system, smart contract plays a major role in the functioning of the system. Smart contracts are self-executing autonomous computer programs that get executed based on the condition defined by the programmer. These contracts are capable of facilitating, enforcing and executing agreements between two parties using blockchain. For example, one can pay room rent automatically at the end of the month without involving a bank in between. Smart contracts have various possible applications such as trading or loaning of properties, stock or bond trading in distributed markets. Moreover, it can be also used for autonomous transparent digital voting system or autonomous digital notary contract system.

In this context, we used smart contract; for single smart contract is used in our proposed solution for the whole system. The smart contract contains a mapping of all the registered IoT devices to the respective users and their associated uploaded data they can manage. Moreover, it contains a list of authenticated end users mapped to uploaded data they are allowed to access. All registration, authentication, access control functionalities are governed in a decentralized manner through the smart contract.

2 Literature Survey

Securing the user entries data is more important, and too retrieval. The user authentication was contributed through many papers for future implementation. Here, let us see brief review about some of the researches which contribute to our context.

uPort may be a stage which gives blockchain-based identity for the client and easy-to-use strategy associated with decentralized applications or services. This stage allows clients to set up a digital identity which can be utilized as client character over numerous cloud administrations without any passwords. It gives full control of delicate client information to the client by permitting client to possess and control their advanced resources, safely and especially uncover their information to counterparts to get computerized benefit. Besides, it permits the client to carefully sign and encrypt records, information, messages and exchanges and to send these over the blockchain network to connect with decentralized applications. It is built on beat of the Ethereum blockchain and has three primary components: smart contract, developer libraries and a portable app. The smart contract has the core logic to issue and recoup client personality. The developer library permits designers to coordinate uPort to the third-party application in order to utilize uPort as identity supplier. The portable app issues the personality to the client and gives identity/key administration interface.

Bit_id is an open-source protocol which permits basic and secure client log into cloud service by confirming the client based on the public key and blockchain network. The verification demonstrates the personality of the client to a service by signing a challenge. A cloud supplier can utilize bit_id to enable blockchain login for their service. This disposes of the requirement of client enrollment and username/password and gives information proprietorship to the client instead of the cloud supplier.

NameID is exploratory innovation based on Namecoin and OpenID to provide unique, secure, decentralized and computerized character to the clients. Namecoin is decentralized Bitcoin-based innovation which permits a client to enlist names which can be related with client data. This information can be verified by everybody within the blockchain arrange but cannot be manufactured or censored by unauthorized assailants, and no one can recover the information without client assent.

OpenID is a protocol that permits a client to confirm to numerous administrations without requiring of making different distinctive personalities and passwords. It provides some special identity to the user from a few trusted identity supplier which can be utilized to sign into other OpenID-enabled administrations. Agreeing to OpenID establishment: over billion clients, counting innovation mammoths like Google, Microsoft, Facebook, Sun, Yahoo! support OpenID empowered client accounts. In this way, NameID combines Namecoin and OpenID where Namecoin issues computerized blockchain identity and relates with client information (username, mail, phone number), and OpenID allows moment sign into over billion of OpenID empowered Web sites.

3 Existing Methodologies

There are various aspects when it comes to exploring security in cyberspace. In particular, we concentrate the security in IoT. There have been various frameworks released for IoT security issues and concerns. But still there is no single industry-based recognized standard accepted by all the firms in general up to date. These frameworks have been released by various top-tier organizations like GSM association, the IoT Security Foundation, the Industrial Internet Consortium.

- PKI and Digital Certificates

Public key infrastructure and the digital certificates play very crucial roles in the development and the security of the IoT devices to the back-end system. They provide the trust and control needed to distribute the public encryption keys and also provide secure data exchanges between user and the device. It also provides proper authentication so as to prevent the snooping in by other users.

- API Security

Application Performance Indicator (API) security is essential to protect the data integrity sent to and from the device back-end systems. It also ensures only the persons with the right level of access can send or retrieve the data.

- Strong Encryption

Strong encryption always plays a major role in the security of any technology while carrying the payload to the device or vice versa. Data at rest and in transmission should be secured using complex cryptographic algorithms. Through which the data are tough to decode even if the level of access is breached.

- Network Security

IoT security also includes the protection of the network in which the device operates. Hence, port security is a major concern, and also disabling port forwarding and using firewalls and blocking malicious IP addresses is also included.

- Network Access Control

The NAC can help identify the IoT devices which are trying to connect to the network and help inventory management of the IoT devices. This provides to track and monitor devices in a network.

4 Proposed Methodology

The main idea of the system using blockchain is mainly for the fact of decentralized approach which provides a new take on the security and authentication issues concerned to the IoT device. With the advent of age comes the advent of technology, security has come a long way from simple hashing techniques to complex cryptographic techniques which store the data and link them based on the previous data. This system aims to reduce the theft of information and also the misuse of any IoT device through wrong access techniques like sniffing. The system uses blockchain as a storage and also as a mechanism in which the records are immune to change by third party or viewers. Using a token-based authentication which resides in the blockchain, we can clearly say that this approach to the security is a new outtake of the conventional methods of encryption of the SSE approach which has various drawbacks starting from time complexity and privacy concerns. This method overcomes all the traditional drawbacks as it is decentralized and does not have a single point of failure unlike the encryption and decryption techniques [3, 4].

In the proposed methodology, the following features for the system were tried to be obtained. The system must have the following user role, namely the admin, the end users, the IoT device. The role of them is as follows.

Admin The admin is responsible for mapping the device to its corresponding user/users at the time of installation of the device in a real-time scenario. An admin can also add one or more admin into the system. He can also add the device mapping or delete the device mapping from the Authentication Contract.

End User The end user is the customer who wants to access the IoT device wired to blockchain. He sends request to the blockchain by specifying the address of the Authentication Contract and invoking the request in the Authentication Contract. His request is stored in the smart contract. He is also linked to a Ethereum Address/Wallet Address.

IoT Device The IoT device is the final endpoint of the blockchain. Once the end user has requested for access to a specific IoT device, the corresponding device gets invoked. The IoT device looks into the authentication contract and checks if the mapping of the user and the device exists, and then if it exists, it retrieves the payload stored by the end user in the contract and deletes the token.

The design of the system is as follows: (Fig. 2).

The data originates from the end user or the person who wishes to authenticate with some particular IoT device. Then, he is requesting for the authentication by invoking methods inside the smart contract which resides on the blockchain. Now, once he invokes the procedure of requesting authentication, a token is created for him and stored inside the blockchain along with the message he wants to pass for the device. Now, once this is done, the particular device is invoked, when it is invoked, it checks inside the smart contract as whether it has received any authentication request from any of its users. If found it recreates the hash and check it with the list of all

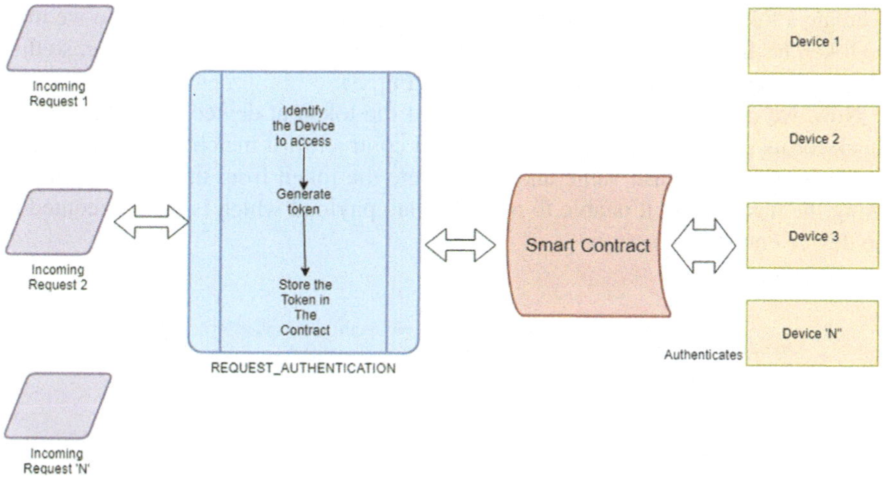

Fig. 2 Design of System

the tokens in the smart contract. If the token matches, then it comes to know that the particular user is authenticating for it, and using the same token, hash retrieves the payload which it so be received by the device. Hence, the authentication is done securely without any mishaps in between. If anywhere the token hash does not match, the transaction is rolled back and no traction is updated in the blockchain so that we can identify only the properly authenticated transaction, and the rest is not possible.

5 Implementation

In this paper, we ensured the user entries list and accessing the users by the help of smart contract with mutual authentication, and we introduced a mapping function for which user accessing which device in the smart contract by the help of user device mapping added by the Admin of the smart contract.

A Etherum smart contract was deployed as a means to hold the user device mapping and also to store the tokens generated by authentication of a device with a IoT device, and finally, we validate the request of the user at the device's end/contract. The proposed system uses a decentralized approach and uses the keccack256 hashing algorithm for creating tokens. We use various block parameters like difficulty and block number to create the hash value of authentication request. The keccack256 is a variant of SHA3 algorithm which is used to securely store the data in the smart contract and also to verify the data in the device end.

This code depicts that it provides the device address which the user wants to authenticate to and also delivers the payload or the instructions for the device to act upon. The address of the device is encoded and user along with the block difficulty to create a token which is to be then verified at the device's end. And, also we map the token hash value with the payload which is to be delivered to the device, so that it can be received at the time of verification (Fig. 3).

Now, we come to the part of validation of the token at device end. The devices check for its user mapping in the main smart contract; if it matches with any of the tokens, retrieve the hash value and then delete the token from the smart Contract. Using the hash value, it is able to map the data payload which is to be executed at the device end (Fig. 4).

```
    function requestAuthentication(address device,string memory
datacode) public {
        // Check if device exists in fog-device mapping
        bool deviceExists=false;
        for(uint256 i = 0; i<users_devices[msg.sender].length;
i++){
            if(users_devices[msg.sender][i]==device){ // check
the devices of a user
                deviceExists=true;
                break;
            }
        }
        if(!deviceExists){
            //trigger DeviceDoesnotExist event
            emit DeviceDoesnotExist(device,msg.sender);
        }
        else{
            emit Authenticated(msg.sender,device);
            bytes32 UID=
keccak256(abi.encodePacked(device,msg.sender,block.difficulty));
            Tokens.push(Token(UID,msg.sender,device));
            data_mapping[UID]=datacode;
            emit TokenCreated(UID,msg.sender,device);
        }
    }
```

Fig. 3 Authentication of a request

```
    function validateTokens()public {
        bool validate=false;

        for(uint256 i = 0 ; i<Tokens.length;i++){
            if(Tokens[i].dev == msg.sender){
                bytes32 tkid =
keccak256(abi.encodePacked(msg.sender,Tokens[i].user,block.diffi
culty));

                if(Tokens[i].UID == tkid){
                    validate=true;
                    emit
SucessfulyRetrievedData(data_mapping[tkid]);
                    delete Tokens[i];
                    emit ValidationSuccess( msg.sender );
                }
            }
          else{
            revert();
          }
        }
    }
```

Fig. 4 Validation of tokens

If the token is not present in the system, the transaction is reverted back using the function revert(). Hence, we validate only the transactions if the device has access from the particular user and also if the user has requested for the access of device from his end.

Due to this fact, we can prevent many middle-man attacks and also impersonation attacks as the ledgers are distributed all over the blockchain, and also, the complexity increases as the blocks increase with the rate of transactions.

6 Results and Discussion

It prevented the data from theft and also achieved mutual authentication for ensuring the user authentication with Ethereum smart contract in model of blockchain. We conclude our process of accessing the database and active nodes list and accessing that nodes are safe over blockchain. Now, let us discuss the results.

6.1 Authentication Contract

The Authentication Contract is the backbone of the proposed system and nearly consists of the entire functionalities inside it. It is a smart contract which is written in solidity a language similar to Java script with its own features. Inside the Authentication Contract, we store the mapping of which device is to be accessed by which user and also the tokens created at the time of request—this token is then mapped to the payload by the requested user and also stored inside the smart contract. This is main module where we use blockchain as a storage for the tokens and authentication and passing of data. The key system inside this contract is the mapping of many-to-many users to device which is contained in a list of address-to-address mapping.

The next part of the system comes to the token generation. The token can be generated based on various attributes; here in our system, we use the user, device and the difficulty of blockchain to generate the token. All these three items are combined together to form a string using ABI Encoders purpose is to combine many items into a single string format. This event log is from Authentication Contract where we can either add admin/user device mappings. Additionally, we see that when a user requests for an authentication, the authenticated event is triggered and the token created event tells us that the token is created and stored in the Authentication Contract (Fig. 5).

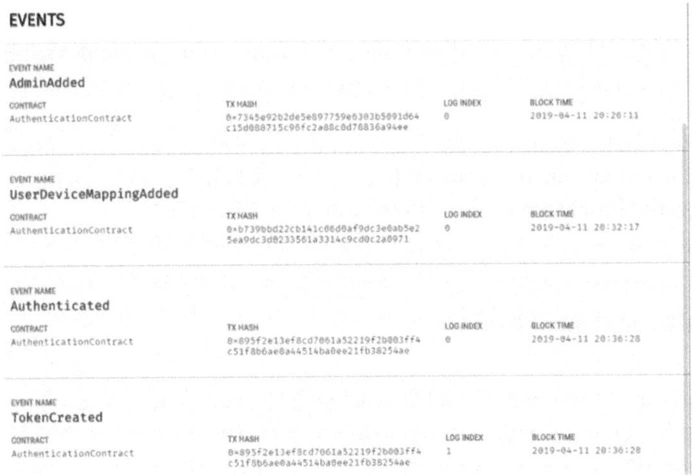

Fig. 5 Authentication Contract events

6.2 User Contract

The user contract is used to form the first step of communication with the blockchain or the Authentication Contract. The user contract is mapped to a wallet address of the end user. The address implicitly invokes the Authentication Contract and communicates with the Authentication Contract and sends it the relevant details like what device which the user is requesting access for and also the data which he wishes to send to the device. The user contract does not hold any data and just merely passes them to the Authentication Contract. If the data are accepted by the Authentication Contract, transaction is recorded on the user contract to prevent falsely accessing data and leading to security leaks.

When the device tries to verify the authentication request done by the user and if the hash matches with the hash stored in the Token array, the device receives the payload sent by the user through the same hash value by which token is created. It is noted that each token's hash value is mapped with its corresponding payload (Fig. 6).

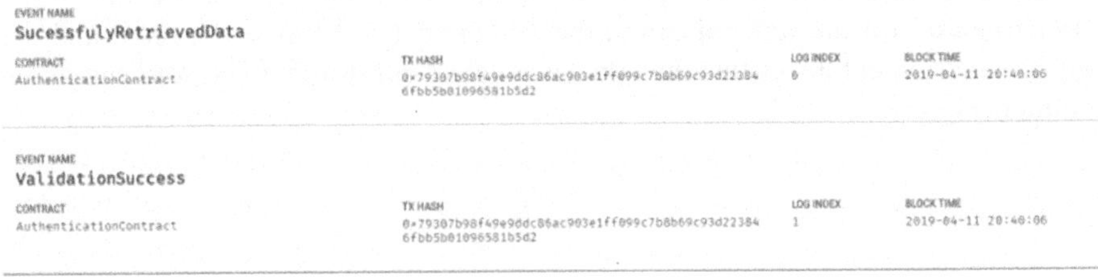

Fig. 6 Authentication successful

The screenshot shows the validation took place successfully in the device end by contacting for details in the Authentication Contract.

6.3 Device Contract

The device contract is very much similar to the user contract except the part that it is linked to the device (IoT). The device contracts also reside on the blockchain and similarly communicate among the contracts and access the Authentication Contract, and hence, it implicitly checks and invokes the *valdiateToken* method in the Authentication Contract and checks if any request is waiting for it using the list of tokens in the Authentication Contract [5]. If such token is there retrieving the payload from the Authentication Contract and executes them in the device end then the execution is done. Once the execution is complete, the transaction is done in the device contract and also the event is triggered in the Authentication Contract, respectively.

7 Conclusion

Through this, we come to know that we can eliminate middle-man attacks as the device mapping is stored inside blockchain, and if anyone impersonating the user is trying to access, the mismatch in user address occurs and the token is not generated; hence, the IoT device is protected from middle-man attacks and impersonation attacks [6]. As we all know that blockchain is decentralized, the transaction is tough to hack or to modify once the data is written to the smart contract [7]. Hence, we can say that this authentication system is more secure than the traditional security which uses encryption and decryption of data alone with a centralized cloud. The future work for the system is to interface with a real-world IoT device and deploy the system in the real world. Also, the future research work would be to make the transactions with a little bit of price so that to create a perfect ecosystem of devices patched up with crypto-currency in the real world.

References

1. Nakamoto S (2009) Bitcoin: a peer-to-peer electronic cash system, pp 10–24
2. Cresitello-Dittmar B (2016) Application of the blockchain for authentication and verification of identity
3. Hummen R, Shafagh H, Raza S, Voig T, Wehrle K (2014) Delegation-based authentication and authorization for the ip-based internet of things. In: 2014 eleventh annual IEEE international conference on Sensing, Communication, and Networking (SECON), pp 284–292

4. Mahalle PN, Prasad NR, Prasad R (2014) Threshold cryptography-based group authentication (tcga) scheme for the internet of things (IoT). In: 2014 4th International Conference on Wireless Communications, pp 1–5
5. Zheng Z, Xie S, Dai H-N, Wang H (2018) Blockchain challenges and opportunities: a survey. Int J Web Grid Serv 14(4)
6. Porambage P, Schmitt C, Kumar P, Gurtov AV, Ylianttila M (2014) Two-phase authentication protocol for wireless sensor networks in distributed IoT applications. In: 2014 IEEE Wireless Communications and Networking Conference (WCNC), pp 2728–2733
7. Underwood S (2016) Blockchain beyond bitcoin. Commun ACM 59(11):15–17. https://doi.org/10.1145/2994581
8. Alonso A, Fernndez F, Marco L, Salvacha J (2017) Iaacaas: Iot application-scoped access control as a service. Future Internet 9(4):64

Detection of Packet Dropping Nodes in Wireless Mesh Networks

R. Thillaikarasi and S. Mary Saira Bhanu

Abstract Wireless mesh network (WMN) is a wireless ad hoc network which uses mesh topology to connect the radio nodes together and it is prone to vulnerabilities due to its open architecture. Packet loss is the major issue due to its open wireless physical media, frequent topological changes, scalability and power constraints. In general, packet drops in WMN during transmission may happen due to mobility or buffer overflow or power depletion or malicious behavior of the intermediate nodes which degrades the performance of WMN. This paper proposes an adaptive dynamic source routing protocol (ADSR) to detect the misbehaving node using the cross-layer approach that performs packet dropping and reroute the packets using the alternate paths. These packet drops are identified precisely in this work, which significantly reduces false positive rate of detection of malicious nodes. The performance of ADSR is analyzed using a mathematical model by considering the characteristics of a node's behavior. Using extensive simulations, the performances are thoroughly analyzed and evaluated and compared with traditional dynamic source routing (DSR), Watchdog (WD), Bait-DSR (BDSR) protocols. Simulation results show that with negligible increase in overhead, this protocol provides secured routes with an increased packet delivery ratio (PDR) and throughput.

Keywords Cross-layer design · Dynamic source routing · Packet drops · Packet salvaging · Wireless mesh network

1 Introduction

With the rapid development of wireless communication technologies, WMN plays an important role because of its reliability and resilient nature. Mesh clients (MC) (user

R. Thillaikarasi (✉) · S. Mary Saira Bhanu
National Institute of Technology Tiruchirappalli, Tiruchirappalli, Tamil Nadu, India
e-mail: thillai-cse@saranathan.ac.in

S. Mary Saira Bhanu
e-mail: msb@nitt.edu

S. M. Thampi et al. (eds.), *Applied Soft Computing and Communication Networks*, Lecture Notes in Networks and Systems 125,
https://doi.org/10.1007/978-981-15-3852-0_3

terminals) and mesh routers (MR) are the components of WMN where MRs form the backbone for WMN through establishing the mesh connectivity between them. Comparing with other multi-hop wireless networking technologies, WMN has sporadic topological changes and it is cost-effective [1]. It is interoperable with existing wireless network technologies like WLAN, WI-FI, Wi-Max, WSN and consequently prop up wide variety of applications. Failure of a node will not affect the communication among other nodes in WMN using mesh connectivity and usage of enhanced routing protocols. Due to its characteristics such as self-healing, self-organizing, minimal cost for installation, and low-cost maintenance, it is widely used in the applications like VOIP, home automation, internet access for communities, and emergency environments like tunnels, battlefield surveillance. Packet drops in WMNs are caused by one or more of the following reasons:

1. Link failure: It may happen in the network due to overutilization of link.
2. Mobility: Mobility of the nodes causes link failures which results in packet loss.
3. Buffer overflow: Because of the shared access medium and the interface queue management used at data link layer of the DSR protocol queue overflow happens frequently which leads to packet drops.
4. Node failure: Nodes may fail to accomplish their task due to energy exhaust during the operations like receiving the data, transmitting the data, overhearing the traffic, and by transmitting control messages RREQ, RREP, RERR.
5. Error in data: Packets are dropped by a node when it detects errors in the packet received.
6. Malicious activity: Intentionally drops the packets forwarded through them and launch denial-of-service attacks (DoS) in the network.

Securing WMN from different types of network layer attacks such as blackhole/sinkhole, grayhole, wormhole, rushing attack, and impersonation attack is a challenging task. Consequently, by embedding security mechanisms to routing protocols or by designing an appropriate intrusion detection technique, these attacks can be detected. Blackhole and grayhole attacks are the most common packet dropping attacks on WMN, and according to the survey done by Jaydip sen [2], most of the routing protocols failed to address these issues. On demand, reactive DSR protocol is suitable for WMN by the way of having multiple routes in route cache which significantly reduces new route discovery process when Route Error (RERR) message is received by the source node during data transmission.

In paper, Akyildiz and Wang [3] have reviewed how cross-layered design plays a crucial role in improving the network performance in an extensive way. Adaptive DSR protocol proposed in this paper makes use of facts collected in MAC layer, physical layer, and these metrics are exploited in the network layer to detect the packet dropping nodes. The appropriate reason for packet drops is identified by checking various parameters like queue length, remaining energy of a node, and mobility nature of a node. Doing so, the legitimate nodes are not erroneously identified as misbehaving nodes which is the shortcoming found in most of the research works such as Watchdog [4] and BDSR [5]. This paper is structured in the following way. Related works are discussed in Sect. 2. Section 3 describes the proposed work. Simulation environment and experimental results for different scenarios are discussed in Sect. 4. Finally, Sect. 5 presents conclusion.

2 Related Works

Most of the researchers have contributed their works to provide security to WMN since securing WMN is a challenging issue which is to be solved. WMN requires two level of security [6]: (i) Securing MCs can be done with client authentication and access control standard techniques. (ii) Backbone security. All MRs build a backbone network using mesh connectivity and they are intelligent devices responsible to exchange the data and control messages. Symmetric and asymmetric encryption techniques are generally proposed by most of the research works to alleviate attacks on network and ensure confidentiality and authenticity of the traffic. Jaydip sen [2] have documented a distinct variety of secured routing protocols with the conclusion that it is not feasible to eradicate blackhole and grayhole attacks completely.

According to Gao and Wei [7], packet dropping nodes are detected by using aggregate signature algorithm. The association between the nodes is mathematically computed as either known or unknown in [8]. A path-based method is proposed by Cai et al. [9] overhears the next hop's behavior. BDSR [5] detects and overcomes the blackhole attacks proactively by using Bait RREQ and tempts the malicious node to reply with RREP. During data transmission, if the data delivery rate is lower than a predefined threshold value (90%), then the destination node sends a message to the source node to trigger proactive process again. But the disadvantage of this method is Bait RREQ might be triggered if packet delivery rate (PDR) decreases due to various other reasons like queue overflow or power depletion experienced by the nodes. Fake RREQ packets are used to detect the malicious nodes in DBA-DSR [10] protocol.

AODV protocol is enhanced by using coming route reply table (CRRT) to store information about the RREP packets along with sequential numbers [11]. The sequential number in the RREP is compared with threshold value, and if it is greater than threshold, then the node that sends RREP is detected as blackhole node. Yuan et al. [12] recommended a technique to identify key node-oriented blackhole attacks based on the behavior of the neighbor nodes. An explore-based active detection (EBAD) scheme on DSR proposed by Pu et al. [13] is used to mitigate routing misbehaviors using fake RREQ messages and fake RREP messages generated by the misbehaving nodes.

In method hybridization of particle swarm optimization with genetic algorithm (HPSO-GA) proposed by Thanuja et al. [14], blackhole attack is detected by using data routing information collected from each neighbor node. Shu et al. [15] proposed a public auditing architecture to validate whether the packets are dropped intentionally or not. Zhang et al. [16] developed a Audit-based Misbehavior Detection (AMD) model to identify and isolate the packet dropping nodes. From the literature review, it is noted that only a few of the prior works have solicited for discriminating the malicious packet drops with normal packet drops due to collision and link failure.

The behavior of the nodes in wireless network can be analyzed using a mathematical model. An explicit stochastic model is proposed in this paper where the node transitions occur according to the transition probability matrix and transition time is a random variable. The effect of correlated failure nodes on network survivability is elaborated by Azni et al. [17] and the effect of mobility in node behavior is discussed in paper [18]. Network Security Risk Assessment Model (NSRAM) [19] is used to detect the vulnerabilities in advance using Semi-Markov model. Most of the research works discussed in this section are used to detect malicious nodes and their influence on network performance. This paper focuses on the actual cause for packet drops in the network and it is also verified using the mathematical model. Simulation results show that with the negligible increase in network overhead, the proposed protocol provides secure routes with increased PDR and throughput.

3 Proposed Work

In this proposed work, all the nodes are assumed to operate in promiscuous mode (They are able to overhear the packets in their transmission range), having a fixed queue length and fixed amount of energy is used to transmit a packet. All nodes make use of Over Hearing Table (OHT) to record information about packets transmitted and received by the nodes in their transmission range. OHT plays an important role in detecting malicious nodes. Nodes also use blocklist array to store IP addresses of the misbehaving nodes. In ADSR prior to commencement of actual transformation, the source node will broadcast RREQ messages to the neighbor nodes with a fictitious destination address to grab the nodes aiming to do blackhole attack. Malevolent nodes are the first one to respond with the RREP message; subsequently, the source node adds the misbehaving node's address in blocklist and broadcasts this information to all other participating nodes in the network.

Eventually, the source node begins to proceed with transmitting the actual RREQ messages and receive RREPs from destination or from intermediate MRs (IMR). After receiving RREP messages, source node selects a path having minimal hop count to reach the destination node and starts the data transmission. Consequently, each and every forwarding MR (F_MR) stores the packet id in forward buffer (FB), forwards the data packet, waits for link-level acknowledgment (LLA) from the next neighbor MR(NN_MR). F_MR promiscuously monitors whether the NN_MR is forwarding the data packet or not. The main cause for not receiving LLA from the NN_MR is either due to mobility of NN_MR or link failure. F_MR makes use of packet salvaging technique to reroute the packets if alternate path resides in its cache or transmit RERR message to the source node to imply that NN_MR is not reachable.

Denial of not forwarding the data packet by NN_MR is due to various reasons like: (i) NN_MR may go out of transmission range due to mobility; (ii) buffer may likely to overflow which might lead to congestion; (iii) the energy level may go beyond the minimum threshold value which might lead the node to shut down; and (iv) NN_MR may be a malicious node that performs packet dropping attacks.

Packet drop due to mobility is identified by F_MR by computing the time difference between sending and receiving of two consecutive packets and their corresponding acknowledgments. The latest time difference is much greater than prior one implies that NN_MR is departing from F_MR.

If the node is static, then F_MR approximately estimates the remaining queue capacity of NN_MR from the observations it recorded in the OHT. Remaining queue capacity of NN_MR is computed by subtracting number of packets received by NN_MR with a maximum queue size and adding the number of packets transmitted by NN_MR. If the remaining queue capacity is equal to 0, then NN_MR drops the packet due to buffer overflow. If the queue capacity is enough to accomplish data transmission without any drop, then residual energy of NN_MR is estimated and checked whether it is sufficient to transfer a packet or not. Energy required to transfer a packet is multiplied with the total number of packets transmitted and the result is subtracted from the initial energy to compute the residual energy of a NN_MR. If the remaining energy of a NN_MR is insufficient to transfer a packet, then it is concluded that the packet is dropped due to energy depletion; otherwise, the node is spotted out as a malicious node.

3.1 Node Behavior Model

In WMN, the nodes may change their behavior from normal to abnormal due to different reasons as stated earlier. Some of the nodes may be selfish or malicious in nature and they try to affect connectivity and availability of network resources. Thus, node behavior model plays a significant role in analyzing the performance of the WMN. In this paper, a stochastic mathematical model is used to detect misbehaving nodes in a distributed way. This model estimates the stochastic properties of MRs, estimates the model parameters, finds transition probability matrix for the nodes, and finally detects the misbehaving nodes. The model is validated using simulation results. According to the behavior of the nodes, states of the nodes are assumed to be as follows:

Legitimate state: The nodes comply with the protocols to be pursued to accomplish data and control message transmissions are in legitimate state.

Failed state: Nodes unable to comply with the rules due to reasons like power depletion, link failure, and buffer overflow are in failed state.

Malicious state: The nodes aimed to drop the data packets during data transmission are categorized as nodes in malicious state. The nodes in malicious state should not transit to any other state, and hence, they should be isolated from the network. The transition from one state to another state is illustrated in Fig. 1.

(i) Estimating the stochastic properties and probabilities:
Each F_MR estimates the properties such as total number of packets received, forwarded by NN_MR, remaining queue length, residual energy and mobility nature of the NN_MR using OHT which stores information of the packets it overhears. A

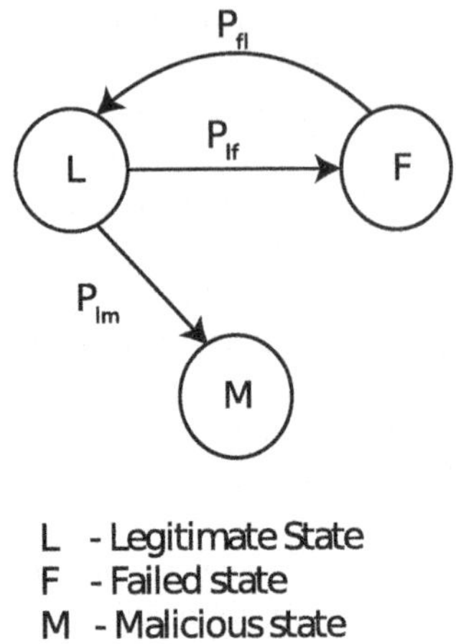

Fig. 1 Node transition model

minimal computational cost in terms of memory and time is required for this. The remaining queue length of the NN_MR is derived using (1),

$$Q_{\text{remain}} = (Q_{\text{Max}} - N_{\text{RP}}) + N_{\text{TP}} \tag{1}$$

where Q_{Max} is the maximum size of the MAC queue and N_{TP}, N_{RP} are the number of packets transmitted and received by NN_MR. Overflow is computed using (2) by checking whether the buffer is full or not.

$$\text{Overflow} = Q_{\text{remain}} \leq 0?1:0 \tag{2}$$

Similarly, residual energy of NN_MR as computed using (3).

$$E_{\text{Res}} = E_{\text{Max}} - (E_{\text{Min}} \times N_{\text{TP}}) \tag{3}$$

where E_{Max} = initial energy of a node, E_{Min} = minimum energy required to transfer a packet, N_{TP} = total number of packets transmitted

Computed E_{Res} is compared with E_{Min} to check whether NN_MR is experiencing energy depletion or not using (4)

$$\text{Energy}_{\text{depletion}} = E_{\text{Res}} < E_{\text{Min}}?1:0 \tag{4}$$

And finally, (5) is used check whether NN_MR is mobile or static by comparing the two consecutive time difference.

$$\text{Mobility} = \Delta 2 > \Delta 1?1:0 \tag{5}$$

where $T1$, $T3$ be times at which F_MR transfers successive data / control packet to NN_MR, $T2$, $T4$ be times at which NN_MR sends LLA to F_MR, $\Delta 1$ is $T1 - T2$ and $\Delta 2$ is $T3 - T4$. Let P_{lf} be the probability of transiting from legitimate state to failed state and is derived from following logical expression (6)

$$P_{\text{lf}} = \text{Overflow}||\text{Energy}_{\text{depletion}}||\text{Mobility} \tag{6}$$

NN_MR fails to forward the packet to its neighbor due to the reasons like power drains below the minimum threshold value or MAC queue may overflow or moves out of transmission range. As a result, $P_{\text{lf}} = 1$ indicates transition of a node from legitimate state to failed state, whereas $P_{\text{lf}} = 0$ indicates no change in the state. Let P_{lm} be the probability of transiting from legitimate state to malicious state and it is defined by the following logical expression (7) and (8) as

$$\text{Forward} = (\text{overheard}P_{\text{id}}\varepsilon FB)?1:0 \tag{7}$$

$$P_{\text{lm}} = \neg(\text{Overflow}||\text{Energy}_{\text{depletion}}||\text{Mobility}) \oplus \text{Forward} \tag{8}$$

$P_{\text{lm}} = 0$ indicates no transition and $P_{\text{lm}} = 1$ indicates a transition from legitimate to malicious state. Further, probability of transition from failed state to legitimate state can be derived using (9) as

$$P_{\text{fl}} = \neg(\text{Overflow}||\text{Energy}_{\text{depletion}}||\text{Mobility}) \tag{9}$$

$P_{\text{fl}} = 1$ means transition from failed state to legitimate state and $P_{\text{fl}} = 0$ means no transition. In this model, we have not considered the transition from malicious state to legitimate state, transition from failed to malicious, and vice versa.

(ii) Formulating Transition Probability Matrix (TPM)

In the Markov process, the time T and state S are discrete and random in nature. The Semi-Markov model $Z = Z_{ij}(t)$ is defined by

$$Z_{ij}(t) = \Pr(S_{(n+1)} = j, T_n \leq t | S_n = i) = P_{ij} \tag{10}$$

where P_{ij} is the transition probability between state i and j. From the stochastic parameters and probabilities derived, TPM is formulated as follows.

$$P = \begin{bmatrix} 0 & P_{\text{lm}} & P_{\text{fl}} \\ P_{\text{fl}} & 0 & * \\ * & * & 0 \end{bmatrix} \tag{11}$$

$P_{ii} = 0$ in P signifies that transition from one state to another state alone is considered and $P_{ii} = *$ means that those transitions are not considered in this model (malicious to legitimate, failed to malicious, and vice versa). Assume NN_MR is having enough resources like MAC queue capacity, battery power but if it fails to forward the data packet forwarded by F_MR, then it will be easily recognized as malicious by F_MR by computing all necessary parameters using (1) to (5) and the probability of transition of NN_MR is predicted using the model. For example: Overflow = 0, $Energy_{depletion} = 0$, Mobility = 0 and Forward = 0 are the values computed for NN_MR, then NN_MR is predicted as malicious node since $P_{lm} = 1$ implies that NN_MR transits from legitimate state to malicious state.

4 Simulation and Results

This section describes the performance analysis of the proposed model and it is evaluated through the network simulator NS3.20. Simulator NS3 is an open-source discrete event simulator which supports different network services, network devices, network protocols, and networking scenarios which fascinate the research community to do research in wired/wireless networks.

4.1 Simulation Environment

The mesh network consists of 64MRs arranged in a 8×8 grid and the initial energy of each node is set as 0.5 J. 10% of the nodes are assumed to be mobile while others are static. Each node is equipped with radio frequency of 802.11b with a data rate of 1 Mbps and the transmission range is set as 500 m. Seven different CBR traffics with distinct source and destination are assumed to be taken place in the network. 512 bytes packets are transmitted with a data rate of 7packets/sec and the simulation time is set as 250 s. The following Table 1 elaborates the simulation parameters used to analyze the proposed model.

4.2 Estimation of Node Behaviors

Node behavior model and its translation matrix formulated in Sect. 3 are verified using various network scenarios in this section.

(i) Effect of energy exhaustion: Every node has initialized with energy $E_{max} = 0.5$ J and their packet forwarding capability will stop once the residual energy E_{Res} goes below to the threshold value $E_{min} = 0.002$J which is the minimum energy required for a node to do a transmission. Node transition due to energy depletion is illustrated in Fig. 2.

During the transmission of 980 packets, energy of a node decreases and E_{Res} computed using Eq. (3) is compared with E_{Min} which results the value of $Energy_{depletion}$

as true and hence P_{lf} as 1. So, the node transits to failed state and drop all packets pass through it.
(ii) Effect of node mobility: Mobility of a node during data transmission also leads to packet loss. To illustrate this the transmission times between two nodes (one node mobile, another static) are recorded and it is observed that for the first three packets Δ value is same and for the fourth packet Δ value is higher than the previous Δ values which clearly shows that node is a mobile node. So, the remaining packets may be dropped due to mobility and hence $P_{lf} = 1$. Behavior of the node changes from legitimate to failed state due to mobility is depicted in Fig. 3.

To illustrate the non-mobility scenario, the transmission times between two static nodes are recorded and noted that the same Δ value for all the packets indicates that there is no mobility and the value of $P_{lf} = 0$.

Table 1 Simulation Environment

Parameters	Values
Simulator	NS 3.20
No.of nodes	64
Topography	3500 m × 3500 m
Simulation time	250 s
Pause time	1 s
Routing protocol	DSR/ADSR/BDSR/WD
Packet size	512 bytes
Transmission range	500 m
Traffic Type	CBR Traffic
Data link type	802.11b
Propagation delay	Constant speed Propagation delay
Mobility	Constant mobility model/Randomway 2D point

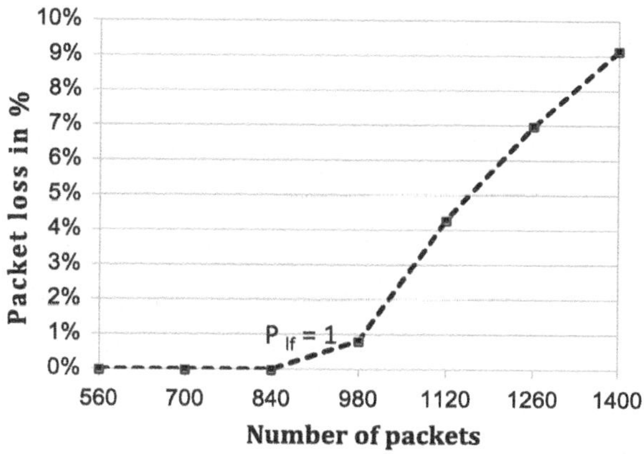

Fig. 2 Node transition from legitimate to failed state due to energy depletion

(iii) Effect of buffer overflow: MAC queue size of all nodes is assigned with maximum value and the total number of packets transmitted over the connections is increased in order to understand the buffer overflow. Q_{remain} are computed using (1), and if it is equal to zero, then overflow flag is set as 1. The above scenario makes the value of P_{lf} as 1, causes node's state transition and packets get dropped due to buffer overflow. Node behavior transition from legitimate to failed state due to buffer overflow is shown in Fig. 4.

The packets transmitted per traffic is increased as 80, 100, 120, 140, 160, 180, and 200, then the total number of packets dropped due to buffer overflow gradually increases and hence PDR decreases.

4.3 Performance Metrics

PDR and network overhead are the metrics used to compare the performance of WD with pathrater, BDSR, DSR, and ADSR with malicious nodes. The effect of false positive on throughput is also evaluated and compared for all the four protocols. The simulation has been conducted with the following setup: (i) All the nodes are initialized with 0.6 J of energy; (ii) queue size is two; (iii) seven connections with CBR traffic; (iv) number of malicious nodes are increased as 6%, 10%, 14%, 20% of 50 nodes; and (v) 10% of nodes are mobile with a speed of 20 s.

Packet Delivery Ratio: It is the ratio between the number of packets successfully received by the receiver and the number of packets transmitted by the sender. 14 out of 64 nodes are used as source and destination to transfer the data packets at a rate of seven packets per sec approximately. The existence of malicious nodes in the network is gradually increased from 6 to 20% of 50 nodes. Figure 5 illustrates %PDR computed for all four methods in the presence of malicious nodes, mobile nodes, nodes with buffer overflow, and nodes suffering from energy depletion.

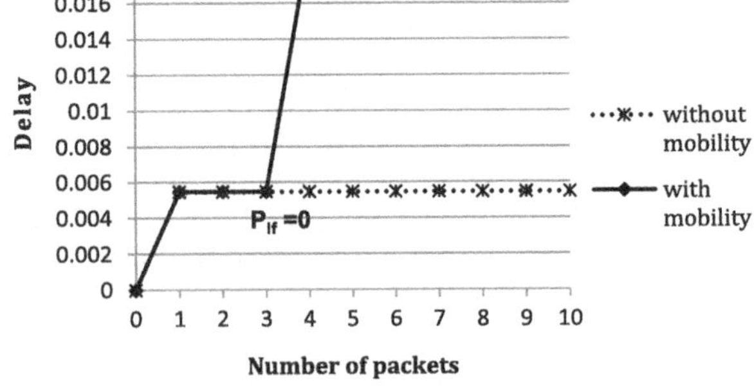

Fig. 3 Node behavior transition from legitimate to failed state due to mobility

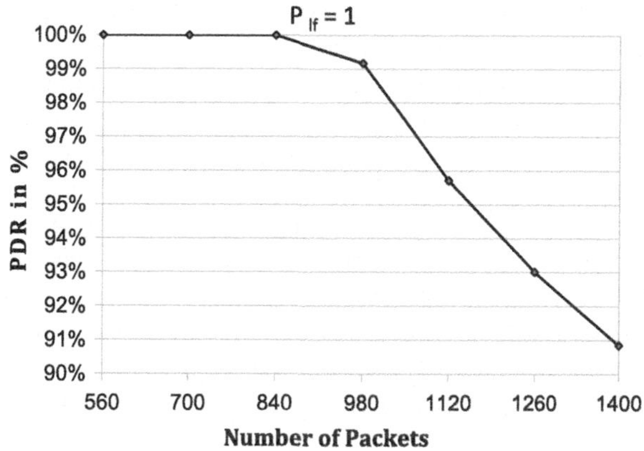

Fig. 4 Node behavior transition from legitimate to failed state due to buffer overflow

PDR of DSR decreases in a drastic way when the number of malicious nodes increases since there is no blackhole detection and prevention mechanism. Watchdog method uses blackhole detection, malicious node isolation technique but the false positive ratio is high. The failure nodes are also falsely isolated as malicious and this results in reduction of PDR. Like Watchdog, BDSR also suffers from a tolerable amount of packet loss as a consequence of increased false positive ratio. But PDR of the protocol proposed in this paper is better than all the other three methods because of decreased false positive ratio.

Detection rate: Apart from detecting malicious nodes in the network, ADSR also detects the actual cause for packet drops in a network. Figure 6 shows how ADSR detects buffer overflow, energy depletion, mobility, and malicious activity when the number of malicious nodes is gradually increased.

Detection rate of malicious nodes is not equal to 100% in all cases due to the non-existence of malicious nodes in the routes used for communication.

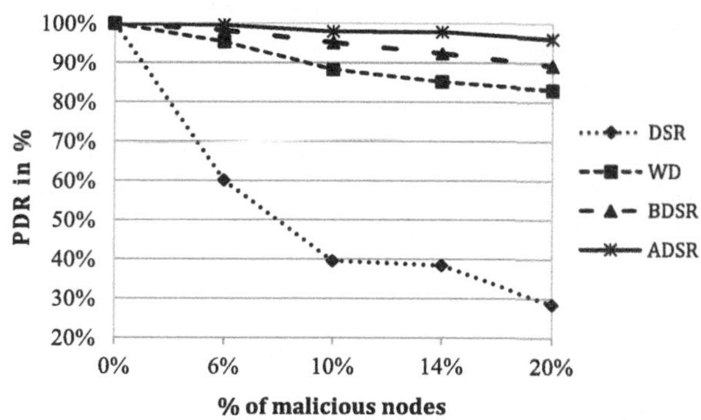

Fig. 5 Comparison of packet delivery ratio

Network overhead: It is a ratio between the number of network-related packets transmitted and the total number of packets (including data packets transmitted). Figure 7 depicts the impact of increased malicious nodes in network overhead for all the four protocols. In Watchdog method, route discovery process is frequently initiated by the false identification of nodes as blackhole nodes and there will be a negligible raise in network overhead. Network overhead of BDSR is comparatively greater than the other methods because dropping off data delivery rate due to mobility or buffer overflow or energy exhaustion of the nodes might lead to the initiation of Bait RREQ which is not appropriate.

The reason for slight increase in network overhead in the proposed model is due to transmission of RERR packets back to the sender when packet salvaging technique fails. Naturally, there should be a tradeoff between network overhead and PDR.

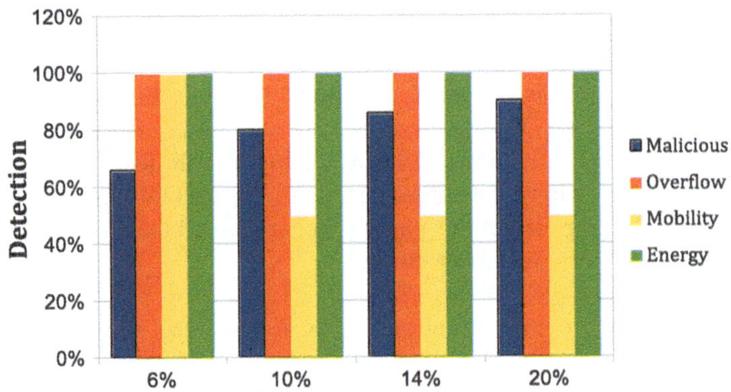

Fig. 6 Detection rate of ADSR

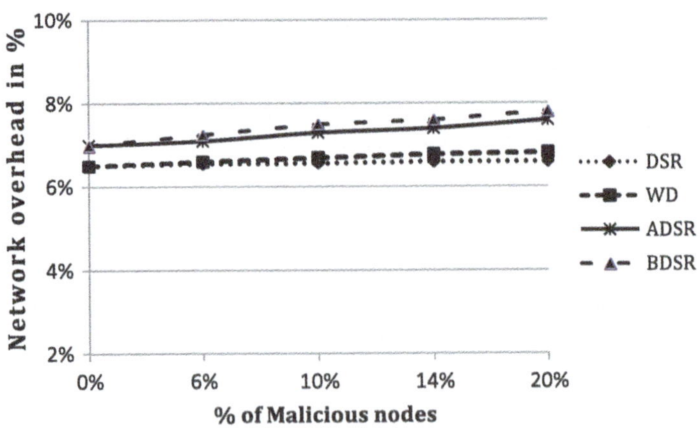

Fig. 7 Comparison of network overhead

Table 2 False positive/True positive of different methods

Protocol	Energy depletion	Buffer overflow	Mobility	Malicious behavior
DSR	Not detected	Not detected	Not detected	Not detected
WD	False positive	False positive	False positive	True positive
BDSR	Not detected	Not detected	Not detected	True positive
ADSR	True positive	True positive	True positive	True positive

False positive: An additional factor vital to compare the different methods is false positive which is the possibility of erroneously identifying a legitimate node as malicious. Due to this behavior, the legitimate nodes are added into the blocklist resulting in reduction of network throughput. The following scenarios are used to evaluate the behavior of the methods considered in this paper, and the results are tabulated in Table 2.
Scenario 1: Energy of some of nodes is at set with minimal value.
Scenario 2: Buffer size of all the nodes is set with a minimal value.
Scenario 3: Some of the nodes are set as mobile and some as static.
Scenario 4: Few randomly chosen nodes in the network are characterized as malicious with packet dropping capability.

The traditional DSR protocol is unable to detect the victim nodes in all the above scenarios because it lacks with detection mechanisms. The WD method erroneously detects a node as a malicious in the first three scenarios and identifies the malicious node properly in the fourth scenario. Whereas BDSR detects the malicious nodes, but failed to detect victim nodes in the first three scenarios like traditional DSR. The proposed ADSR method detects the nodes having (i) depleted energy, (ii) buffer overflow, (iii) mobility, and (iv) malicious activity using appropriate techniques. Furthermore, ADSR reroutes the remaining packets using packet salvaging technique which improves the network performance in terms of PDR and throughput.

5 Conclusion

WMN has extensive future in different areas, e.g., disaster management, military, intelligent transportation system, public safety, monitoring, and home automation. Due to characteristics of WMN such as the open physical medium, lack of centralized monitor and dynamic topology, securing WMN is a challenging issue which is to be solved. Most of the existing solutions discussed in Sect. 2 considered AODV as a routing protocol. This paper discusses an adaptive DSR protocol with minimum routing overhead to overcome blackhole attacks in WMNs. In ADSR, nodes are operated in promiscuous mode to monitor the neighbor's behavior and it can rapidly detect and isolate the nodes performing blackhole attacks. The mathematical model has been used in this paper to study about the node's behavior and nodes change their behavior according to the transition probability matrix. Further, the results of simulation demonstrate that ADSR is better than the traditional DSR, WD, and BDSR protocols while comparing PDR, throughput and network overhead; the detection rate implies that how efficiently ADSR detects the root cause for packet drops. The benefit of proposed work compared with the existing works is it is not necessary to do any complex computations. Detection of cooperative blackhole attacks is the future extension of this work.

References

1. Akyildiz IF, Wang X, Wang W (2005) Wireless mesh network: a survey. Comput Netw ISDN Syst 47(4):445–487
2. J Sen (2013) Security and privacy issues in wireless mesh networks: a survey. In: Wireless networks and security: issues, challenges and research trends, pp 189–272
3. Akyildiz IF, Wang X (2008) Cross-layer design in wireless mesh networks. IEEE Trans Veh Technol 57(2)
4. Marti S, Giuli TJ, Lai K, Baker M (2000) Mitigating routing misbehavior in mobile adhoc networks. In: Proceedings of the 6th annual international conference on mobile computing and networking (MobiCom), pp 255–265
5. Tsou PC, Chang JM, Lin YH, Chao HC, Chen JL (2011) Developing a BDSR scheme to avoid blackhole attack based on proactive and reactive architecture in MANETs. In: 13th International Conference on Advanced Communication Technology (ICACT2011), Seoul, pp 755–760
6. Stallings W (2006) Cryptography and network security, 4th edn. Prentice Hall
7. Gao X, Wei C (2007) A novel grayhole attack detection scheme for mobile Ad-Hoc networks. In: IFIP international conference on network and parallel computing workshops, pp 209–214
8. Bhalaji N, Shanmugam A (2009) Association between nodes to combat blackhole attack in DSR based MANET. 978-1-4244-3474-9/09
9. Cai J, Yi P, Chen J, Wang Z, Liu N (2010) An adaptive approach to detecting black and grayhole attacks in AdHoc network. In: 24th IEEE international conference on advanced information networking and applications
10. Woungang I, Dhurandher SK, Peddi RD, Obaidat MS (2012) Detecting blackhole attacks on DSR-based mobile adhoc networks. In: 2012 international conference on Computer, Information and Telecommunication Systems (CITS). Amman, pp 1–5. https://doi.org/10.1109/CITS.2012
11. Kumar V, Kumar R (2015) Adaptive approach for detection of blackhole attack in mobile Ad hoc network. In: International conference on intelligent computing, communication & convergence. https://doi.org/10.1016/j.procs.2015.04.122
12. Ding Y-M, Wu H (2016) Key node oriented black hole attack and detection. In: Joint international conference on Artificial Intelligence and Computer Engineering (AICE 2016) and international conference on Network and Communication Security. https://doi.org/10.12783/dtcse/aice-ncs2016/5719
13. Pu C, Lim S, Chae J, Jung B (2017) Active detection in mitigating routing misbehavior for MANETs. Wirel Netw. https://doi.org/10.1007/s11276-017-1621-z
14. Thanuja R, Umamakeswari A (2018) Black hole detection using evolutionary algorithm for IDS/IPS in MANETs. Cluster Comput. 1–13. https://doi.org/10.1007/s10586-018-2006-5
15. Shu T, Krunz M (2012) Detection of malicious packet dropping in wireless Adhoc networks based on privacy-preserving public auditing. In: Proceedings of WiSec, pp 87–98
16. Zhang Y, Lazos L, Kozma W (2016) AMD: audit-based misbehavior detection in wireless Ad Hoc networks. IEEE Trans Mobile Comput 15(8):1893–1907. https://doi.org/10.1109/TMC.2012.257

17. Azni AH, Ahmad R, Noh ZAM, Basari ASH, Hussin B (2012) Correlated node behavior model based on semi Markov process for MANETS. Int J Comput Sci Issues (IJCSI) 9(1)
18. Zhao M, Wang W (2006) A novel Semi-Markov smooth mobility model for mobile Ad hoc networks. In: Proceedings of IEEE GLOBECOM, pp 1–5
19. Sun F, Pi J, Lv J, Cao T (2017) Network security risk assessment system based on attack graph and Markov chain. J Phys

An Enhanced Bat Algorithm for Parallel Localization Based on a Mobile Beacon Sensor in Wireless Sensor Networks

Miloud Mihoubi, Abdellatif Rahmoun, and Pascal Lorenz

Abstract Wireless sensor networks (WSNs) have done extensive work over the past decade, focusing primarily on node position in large WSNs when using Internet of Things. When a few anchor nodes are established, the location of sensor nodes must be found and defined in a large network. The goal is to solve this problem by means of an energy-minimal and high-precision protocol. We suggest an improved parallel position scheme based on Bat algorithms, in this article. The main idea is to use a mobile node in the middle of certain network nodes. Wherein, the center node moves along a spiral path to locate its neighboring nodes in parallel. Simulation results of some case studies show remarkable localization quality with regard to accuracy and energy consumption.

Keywords Parallel localization · Wireless sensor network · Optimization · Bat algorithm · Exploration · Exploitation · Internet of Things

1 Introduction

The wireless sensor network (WSN) on the Internet of Things (IoT) has received tremendous attention and earned considerable coverage in the last decade. The scope of WSN implementations has covered different applications such as storage, biological control, and mobile computing [1]. WSN consists of a large number of randomly deployed sensor nodes in the focus area. Nodes shall be liable to detect, store,

---

M. Mihoubi
EEDIS Laboratory, University of Djillali Liabes, Sidi Bel-Abbes, Algeria
e-mail: mil.mihoubi@gmail.com

University of Ibn Khaldoun, Tiaret, Algeria

M. Mihoubi · A. Rahmoun
LabRI-SBA Laboratory, High School of Computer Sciences, Sidi Bel Abbès, Algeria
e-mail: a.rahmoun@esi-sba.dz

P. Lorenz (✉)
IRIMAS Laboratory, University of Haute Alsace, Colmar, France
e-mail: lorenz@ieee.org

S. M. Thampi et al. (eds.), *Applied Soft Computing and Communication Networks*, Lecture Notes in Networks and Systems 125,
https://doi.org/10.1007/978-981-15-3852-0_4

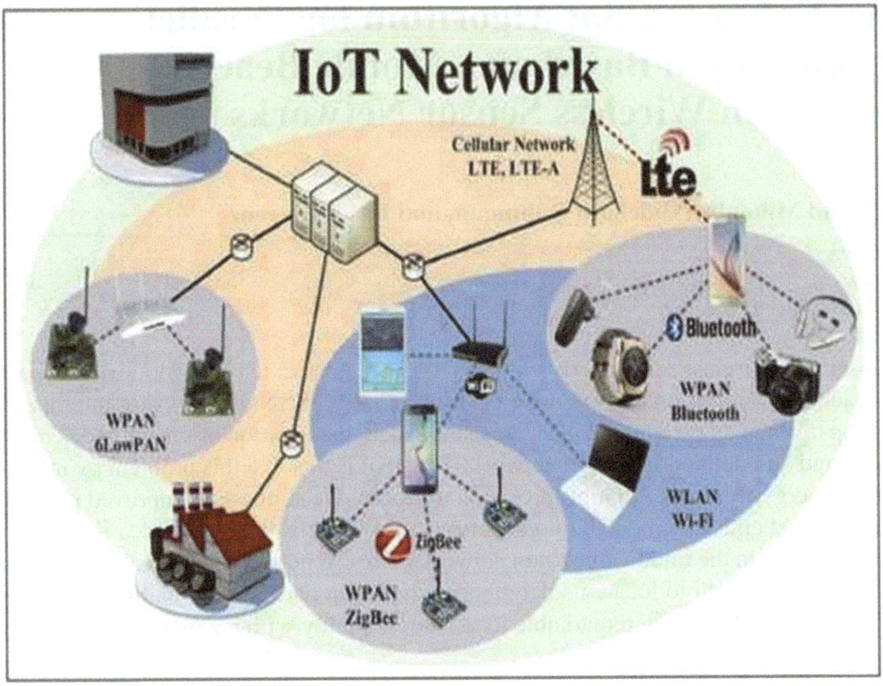

Fig. 1 IoT communication architecture

and transfer the gathered data to baseline station through a many leaps, and the communication architecture is presented in Fig. 1.

WSNs have been enormously investigated in many application domains such as monitoring and security functions.

Sensor networks have been deployed to support surveillance capabilities such as threat-presence detection within security-sensitive and hostile regions such as a militarized area and border protection. To support the monitoring project, a WSN was proposed in [2] to monitor, control, and define the movement direction of invading peoples and the intelligent machine in the sensitive field of application.

Healthcare services are also being referred to as wearable and implantable body sensors linked to the popular body sensor network (BSN). A BSN design consists of a series of thumbnails, lightweight, low-energy, capture tools, and wireless transmitters. These kinds of networks are aiming to collect appropriate patient health data, and the gathered data allows remote monitoring of the patient and enable the doctor to instantly interfere in emergency situations [3]. One major concern of WSN is sensor localization. Calculating the exact coordinates of unidentified sensors nodes is the main objective of localization [4]. The localization operates the actual coordinates of landmarks sensors and intersensor processing. The node has location information called landmark or beacon node, whereas the node has no location information called an unknown node.

Fig. 2 Internet of Things applications

For several monitoring applications, adequate localization adaptability is strongly necessary, for example, farming networks monitoring, inventory management, monitoring of water quality, target tracking, and mortar launch; see Fig. 2.

Every sensor node can be fitted with a Global Positioning System (GPS); clearly, the solution proposed is unacceptable for two chief considerations:

- The expensive machine in the term: energy, covered area and field.
- In specific environments (underwater, buildings, etc.) work with difficulty.

The new trend to deal with the node localization problem is based on minority landmark nodes along with transmission techniques [5, 6]. Alternatively, these depend on unknown node localization to connect each sensor node to a GPS device.

In addition, the node position issue has been shown to be a complex problem, so swarms are tackled to solve the location question [7]. Recently, Bat algorithm was proposed to tackle complicated problems.

In the exploitation portion, the Bat algorithm is proven to be effective, yet the exploration part is poor. In this paper, an improved Bat algorithm for localization is introduced, and to conserve energy consumption in the localization process, the changing rate is recommended in Bat. In addition, the intensity variable is recommended to be modified at each iteration and parallel localization depends upon the wireless beacon which is also proposed.

The paper is divided into seven pieces. Part 2 gives an overview of metaheuristic algorithms which used to locate nodes and also presents the corresponding work. Part 3 provides a Bat algorithm description. The enhanced approach is outlined in Part 4. Part 5 summarizes a simulation study under certain parameters of assessment. Our conclusion and future work are drawn up in the last section.

2 Stat of the Art

The localization of nodes has become a major issue in recent years, and a lot of researches have outlined the issue [8, 9]. The localization category is classified according to various criteria many parameters: mobile or static node, range-based or range-free, distributed or decentralized environment [10].

The node localization challenge has been researched extensively; initially, mathematical methods have been studied tremendously and explored extensively.

In [11], maximum likelihood is suggested for node location and the problems lead to a nonlinear optimization problem; in [12], Distributed Weighted Multidimensional Scaling (DWMS) is proposed as a solution to deal with node localization challenge, and the main idea is that every node selects its neighbors, updates the regional value functions, and uses the same technique to link its neighbors. Inefficiency in overcoming the localization task has been shown in quantitative techniques since several requirements are not taken into account. Recent approaches were focused on swarm intelligence techniques.

2.1 Centralized Technique

Particle swarm optimization (PSO) is intended for the treatment of node localization problem [13], and the objective is to find the target nodes using a few landmarks; thereafter, each target node passes its coordinates and its distance to the base station. The latter separately performs the PSO algorithm to test the objective variable as a mean square error; furthermore, the proposed scheme has some drawback in the hierarchical technique:

(1) The aspect with flip uncertainty is neglected. (2) That node interacts directly in a hierarchical network with a base station, so that high connectivity is required. (3) That sensor node's approximation array is controlled by base stations, which generate a large amount of communication, so that the network becomes vulnerable to energy depletion.

2.2 Decentralized Technique

A single hop range is used in the localization of WSN in distributed techniques; the location of an unknown node may be estimated where it has at least three landmark nodes. It is stated that this node is a node located. The distance from each localized node is measured by each of the anchors around it.

Differential evolution (DE) is proposed to deal with localization [14]; each node executes independently algorithm DE to calculate its positions, the simulation result has proved that the algorithm converges rapidly after a few iterations (22), flower pollination algorithm (FPA) is proposed for localization issue [15], and simulation results show the efficiency of the FPA algorithm, in which several parameters for PSO variant algorithms are examined. The three most popular metaheuristic algorithms (Bat, PSO, and DE) are compared to each other in [16], an effective Bat algorithm is proposed by [17], and further, the drawback of Bat algorithm is overcame by introducing Doppler effect phenomena. The updating frequency makes the algorithm converge rapidly during the Bats' motion and localization error is minimized, in addition. Gray wolf optimization (GWO) is another algorithm proposed to deal with node localization problem [18], GWO mimics based on four types of gray smart wolves to localize and hunting prey, and GWO proved that it has faster convergence rate estimating the location of the node, along with the minimal computation time.

2.3 Multi-stage Technique

Elephant herding optimization algorithm (EHOA) in [19], moth flame optimization algorithm (MFOA) [20] and whale optimization approach are the latest metaheuristic approaches proposed to deal with node localization in a distributed paradigm of WSN [21]. The crucial idea of the technique is based on multi-stage localization; multi-stage localization is based on cumulative nodes, i.e., the localized node in current iteration will become as beacon nodes in upcoming iteration, and this strategy makes the amount of beacons nodes increases at each iteration, and at the same time, localization computing decreases and also localization error decreases.

2.4 Mobility Technique

This technique is based on few mobile anchor nodes (MANs) to localize the remains unknowns nodes, MANs are equipped with GPS units, MANs move in the network along a defined path and periodically broadcasting their actual locations to aid the unknown neighbor nodes, and a significant work has been conducted. In [22], equilateral triple optimal coverage is proposed to calculate the actual coordinates of the virtual anchor's nodes, ant colony algorithm (ACA) is proposed as a guide to determinate the best coordinates of the unknowns nodes, and the beacon nodes browse all the virtual anchor's nodes by running ACA.

In [23], an enhanced PSO is proposed as a solution to calculate the best positions of unknown nodes in a square environment, a spiral path is adopted to browse the network starting from the axe of the field, the global worst particle is used to update the velocity of PSO agents, and the obtained results are very effective in terms of precision and computing.

A mobile anchor node technique is presented in [24], one mobile anchor station and received signal strength indicator (RSSI) technique are combined to localize the unknowns nodes, and the obtained simulation and the comparative study conducted shown that localization error is reduced and the computing time is minimal.

In [25], for underwater acoustic sensor networks (UASNs) a remote node position algorithm focused on compressive sensing is introduced, the sub-water monitoring field cubic unit is segregated into sectors although energy consumption is regulated by the interplay tracking process, and the fragmented cubic unit is transformed into some kind of compression sensor-dependent. In addition, the mobile nodes browse randomly the entire regions of the area. The proposed system conserves the network energy while increasing the accuracy of the location.

A new paradigm trajectory is designed for mobile landmark localization labeled M-curves [26], i.e., the mobile node pursues the "M" design; therefore, dolphin swarm optimization (DSO) has been applied to enhance localization precision, the proposed path provides extensive coverage and strong position reliability relative to other compared static designs LMAT, Z-curves, HILBERT SCAN. Nonetheless, it indicated that energy consumption was completely disregarded. That did unreasonably; browsing the entire network through M-curves path allows exhausting the mobile node and therefore breaks down the network.

3 Bat Algorithm (BAT)

Bat algorithm is a novel metaheuristic population-based algorithm developed by Yang [7], it is inspired by the echolocation behavior of fascinating animals named bats, the micro-bats utilize an echolocation to finds its prey, locate their roosting crevices in the darkness, and avoid obstacles; this micro-bat has the ability to find out the coordinates of prey by using forwarding high signal and short audio signals. The advanced echolocation capability of Bats makes them fascinating and works perfectly in dark environment. Substantially, the main five steps of Bat algorithm are:

Step 1: Initialization of Bat Population

In the first place, set of Bat "N" is distributed over the search field without any information about the place of the prey; however, the groups of the population are indiscriminately generated for each dimension value "*d*" with initial account for loudness, velocity frequency and pulse emission rate

$$x(i, j) = \mathrm{LL}(j) + rand(0, 1) \cdot \mathrm{UL}(j) - \mathrm{LL}(j) \quad (1)$$

Whether $rand \in [0, 1]$ is a random vector drawn from a uniform distribution, $i = 1, 2 \ldots N, j = 1, 2 \ldots d$, the value of upper limit (UL) and lower limit (LL) depends on the network area (X_i, Y_i).

Step 2: Movement of Virtual Bats (update frequency, velocity, and position)
Each population Bat generates spontaneous-frequency sound signals, which regulate acceleration and update (new solution) positions, and modern velocity is controlled at phase t as follows:

$$f_i = f_{\min} + (f_{\max} - f_{\min}) \cdot rand \quad (2)$$

$$v_i^{t+1} = v_i^t + (x_i^t - x^*) \cdot f_i \quad (3)$$

$$x_i^{t+1} = x_i^t + v_i^{t+1} \quad (4)$$

where $f_i \in [f_{\min}, f_{\max}]$ is frequency, x_i^t is position, v_i^t is velocity, $rand \in [0, 1]$ is a random vector drawn from a uniform distribution, and x_* is the actual global best candidate solution.

Step 3: Local Search
The Bat candidates' team query of local search regions only produces one solution, while the following formula is used to find a new solution dependent on the random walk:

$$\mathrm{if}(rand(0, 1) > r_i \quad (5)$$

$$x_{\mathrm{new}} = x_{\mathrm{old}} + \varepsilon + A^t \quad (6)$$

where $\varepsilon \in [-1, 1]$ is a random vector drawn from a uniform distribution, A^t represents the medium intensity of Bats participants at step t, $r_i \in [0, 1]$ is the impulsion ratio. When the best value got via the set of N bats are great than the previous function $f(x_*)$, at this time the global best solution x_* can be updated.

Step 4: Updating Loudness and Pulse Emission Rate
If the new solutions are updated, so the loudness and emission rates decrease and increase, respectively; this means that these bats are headed for their prey.

This relationship can be explained as follows:

$$\mathrm{if}(rand(0, 1) < A_i^t \&\& f(x_i) < f(x) \quad (7)$$

$$f(x) = f(x_i) \tag{8}$$

$$A_i^{t+1} = \alpha \cdot A_i^t \tag{9}$$

$$r_i^{t+1} = r_i^0 \cdot (1 - e^{-yt}) \tag{10}$$

where $\alpha \in [0, 1]$ and y $(y > 0)$ are constants, r_i^0 are a random value, $r_i^0 \in [0, 1]$, $\left(A_i^t \in [1, 2] \text{ at } t = 0\right)$.

The summarized flowchart of Bat algorithm is presented by Fig. 3.

Step 5: Search the Actual Best Candidate

Select the best solution between the last best fitness value and improved bat fitness; if it is better, introduce the best bat solution. When the last iteration is reached, the current best bat x_* is upgraded and operates as an issue solving soon.

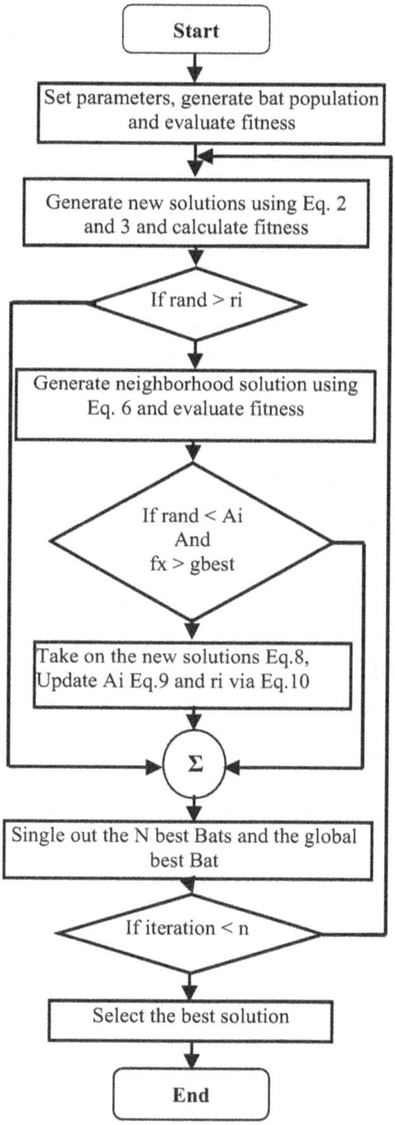

Fig. 3 Flowchart of Bat algorithm

4 Enhanced Bat Algorithm (EBat)

This section modifies the key parameters of the Bat to improve Bat's convergence and speed up Bat movement during localization.

5 Exploration Phase Regularized

The key parameters of Bat convergence are loudness A_i^t and pulse emission rate r_i, where during the increasing of iteration number, we found that pulse emission rate reduces; see Fig. 4.

Figure 5 provides pictorial representation of iteration number versus the pulse rate of original Bat algorithm; from Fig. 5, we show that when the iteration number increases, loudness raises exponentially.

The inverse relationship between the increase of the loudness and the decrease of the pulse rate is related to the principal's phases of the algorithm that are exploration and exploitation, where the exploitation task is exploring a bounding area of the search space based the one-hope technique of improving a promising solution K that we already have at hand. This operation plays a vital role to intensify (refine) the search in the environs of K, by cons, the exploration assignment serves to explore a much larger portion of the search area through the hope linking technique; in addition, take into consideration, finding other hopeful solutions is not yet been elaborated.

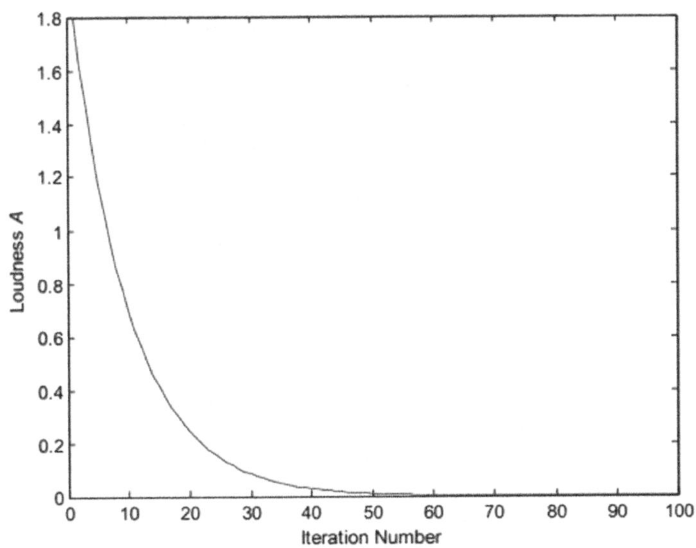

Fig. 4 Iteration number versus loudness

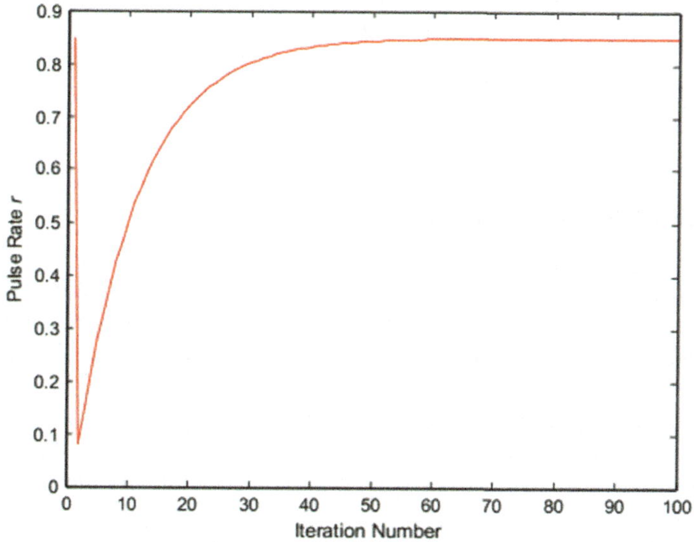

Fig. 5 Iteration number versus pulse rate

This operation amounts to diversifying the search area in order to avoid getting trapped in a local optimum candidate.

Bat algorithm is very efficient in exploitation phase, but it is very ineffective in exploration phase, the equation of the exploitation is selecting the best solutions, if new candidate (solutions) is generated by initialization parameters, the algorithm will be very performing at exploration part but weak at exploitation part; in addition, the solution is feasible in case of the selected best solutions is being generated around the best candidate groups, and the algorithm is weak at exploration part but very efficient at exploitation part. This disadvantage makes the algorithm less efficient of optimization, for making more performance of the algorithm, we have balanced between exploration and exploitation phases, and the Eq. 6 is modified by:

$$x_{ij}^{t+1} = \begin{cases} x_{ij}^{t} + \varepsilon \cdot \bar{A}_{j}^{t} \ldots rand_j > r_{ij} \\ x_{ij}^{t} \ldots \text{otherwise} \end{cases} \tag{11}$$

where u indicates a best-selected solution and $\bar{A}_j^t$ indicates medium loudness of dimension j of the whole solutions at time t. On the other hand, the whole dimensions of any solution check the condition are contained to Bat population and are replaced by the medium loudness as $(\bar{A}_i > rand)$. The new equation of loudness and pulse emission rate is:

$$A_{ij}^{t+1} = \begin{cases} \alpha \cdot A_{ij}^{t} \ldots rand_j > r_{ij} \\ A_{ij}^{t} \ldots \text{otherwise} \end{cases} \tag{12}$$

$$r_{ij}^{t+1} = \begin{cases} r_i^0 \cdot (1 - e^{-yt}) \ldots rand_j > r_{ij} \\ r_{ij}^t \ldots \text{otherwise} \end{cases} \quad (13)$$

6 Parallel Localization Scheme

In this scheme, the five anchor nodes participate on parallel localization; see Fig. 6. Green shape represents the beacons nodes and the bleu shape represents the unknown nodes, where the corner nodes are static, whereas the middle one is mobile and follows spiral trajectory; see Fig. 7.

From Fig. 7, at each one round the beacon mobile node sleeps four times, i.e., sleep, during the movement but do not broadcast any RSSI signal; this strategy is adopted to conserve the energy of the mobile and accelerate the movement during the localization.

The parallel localization algorithm is summarized as follows:

```
Begin
Define topology of WSN.
```

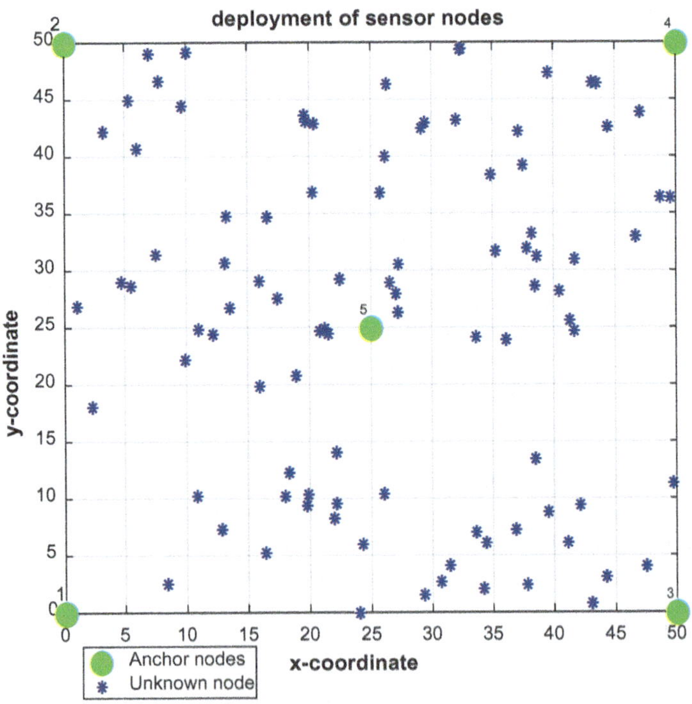

Fig. 6 Sensor node deployment

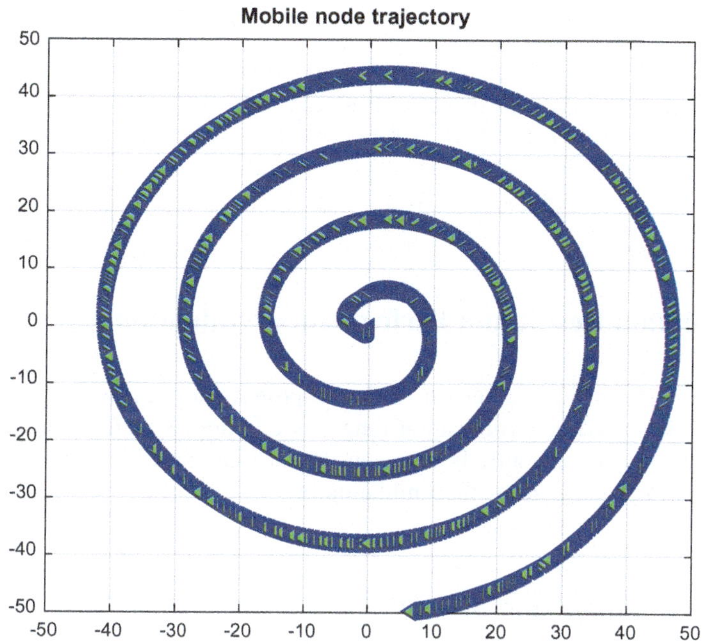

Fig. 7 Mobile node movement

```
{
Initialize Dimension of networks (length and width).
Initialize number of sensors nodes 100   (N) .
Initialize number of Anchor nodes 5   (S)   .
Initialize number of Unknown nodes M=95   (M )
Initialize the communication range.
J=0;
}
While (J<M)
{
Start Middle Node (MN) via a spiral trajectory
Broadcasting RSSI to estimate the distance between the beacon
   nodes and each sensor
Sleep (MN) (moving without RSSI sending)
If neighboring_anchor_n>=3
{ Call multi-lateration to determinate sensor nodes
   coordinate's.
  Call (Algorithm localization):
   (i) Input: (initialize the parameters)
   (ii) Call the objective function f(x,y)
   (iii) Output:
   1- Return the coordinates (x, y).
   2- Plot the coordinates on the figure.
    }
N=N+1; (new Anchors node Added).
J=J+1.
```

```
Else     {  J=J+1  }
End if  }
End While.
Calculate the localization error (Eq. 14) .
Performance evaluation of localization algorithm via several
   criteria
End.
```

7 Simulation Result and Performance Evaluation

Firstly, the performance evaluation of the proposed algorithm is evaluated under iteration number and communication range; in addition, a parallelism scheme is implemented based on mobile beacon sensor and four static beacon also, and the same network is kept to throw the simulation.

7.1 The Fundamental Metric

Two major metrics [27] are suggested as metrics evaluation:

- Average localization errors (ALE): The average distance between the estimated coordinates (x_i, y_i) and the actual coordinates (X_i, Y_i).

$$E_S = \frac{1}{M_S} \sum_{i=1}^{S} \sqrt{(x_i - X_i)^2 + (y_i - Y_i)^2} \tag{14}$$

where "$M = N - S$", M is the number of localized node.

Average localization time (ALT): is the average time needed to localize the entire unknown nodes, ALT is defined as

$$\text{ALT} = \frac{\sum_{i=1}^{N} \text{excution time}}{N} \tag{15}$$

Energy consumption (EC): is the amount of energy used by each mobile beacon a node during localization scheme, EC is defined as:

$$\text{EC} = \frac{\sum_{i=1}^{N} E_{\text{i}}}{N} \tag{16}$$

in which N corresponds to the landmark nodes and E_{i} is really the energy requirements of the landmark sensor nodes active in the localization process. Based on the

Table 1 Enhanced Bat algorithm parameters

Parameters of Bat	Parameter values
Population size Alpha value	10 0.9
Loudness value	0.99–2
Pulse rate value	1
Dimension of the network	2
Sigma value	0.9

current node size and spatial measurements, the preliminary landmark node power is fixed to 1000 J. In addition, Table 1 outlines the parameters of the envisaged method.

7.2 Simulation Setup

In this subsection, all the simulations are performed using Python programming language on a laptop i5-2.4 GHz CPU to evaluate the performance of parallel localization. In simulation experiments, the whole sensor nodes (unknowns and landmarks) are randomly deployed in two-dimensional square area with length of the side equals 50 and based on a continuous uniform distribution pseudo-random generator. The total number of sensor nodes is fixed at one hundred sensors, the landmark rate is fixed at 0.0500%, through five anchors nodes, and the landmark transmission range is varied in the interval [05 m–50 m]. A pattern of node deployment is pinpointed in Fig. 7.

8 Performance Evaluation

Parallel localization scheme is square shape, the enhanced Bat algorithm is evaluated under the proposed scheme, and on the other hand, the proposed algorithm is evaluated under two crucial parameters (iteration number and communication range).

8.1 Localization Results Under Iteration Number

In this subsection, several simulations are carried out. The iteration number is varied as follows: 20, 30, 40, 50, 60, 70, 80, and 90 of all sensor nodes (100 nodes) in addition; the communication range of each sensor node is fixed at 30 m. From Table 2, we show that the first value of iteration number is 20, the Average Localization Error (ALE) is around one meter, from the table, where the initial energy is set at 1000, we show that ALE decreases gradually, respectively, to the growing of iteration

Table 2 Iteration number versus localization results

Iteration number	ALE	ALT	EC
20	1.064	11.8723	960.50
30	0.9244	17.0741	929.20
40	0.7855	23.0465	910.17
50	0.6258	27.9357	905.00
60	0.5653	32.617	901.10
70	0.4851	35.2179	888.32
80	0.4366	49.299	880.00
90	0.3610	41.6792	875.04

number, when node grows, the network relationship between sensor nodes becomes well attached with high connectivity and the number of localized nodes grows, this growth is realized by parallel paradigm applied, also, at each iteration, the new localized nodes at actual iteration will be considered as anchors nodes at upcoming iteration. On the other hand, more the iteration number increases more the localized nodes increase, and the localized area decreases because the additional beacon nodes cover the region of the settled nodes.

The variations of average localization time (ALT) versus iteration number is presented in column two of Table 2; we remark that computing time reduces, respectively, with each augment of iteration number; however, the growth was due to two factors; at each step, the sensor node tries to detect the best candidate in the region, and the localized nodes at actual iteration will be considered as an anchor node in upcoming iteration and effectively participate the next localization operation. The variations of energy consumption (EC) versus iteration number is presented in column three of Table 2; the first initial beacon node is fixed at 1000 J, and in the beginning, we show that there are five beacon nodes; from Table 2, we show that energy is decreased with a rise in iteration, this is due that at each iteration, the beacon node used energy to send manipulate and detect region of the unknown node; in addition, the energy deference absorbed from one iteration to the other is reduced; this is due that the mobile node is placed into synchronous condition after every unknown node detection process. On the other hand, the EC is slowly reduced between iteration 60 and 90, because with every iteration I, the number of located nodes increases and participates in the upcoming iteration as anchor node.

8.2 Localization Results Under Communication Range

The variations of average localization error (ALE) versus communication range is presented in Table 3, the number of beacon nodes is fixed at 5, and firstly, the initial value of communication range is fixed at 50, the obtained ALT is around 48, the unknown nodes are randomly deployed so they influence the distribution of anchors

Table 3 Communication range versus localization results

Communication range	ALE	ALT	EC
50	1.1290	47.1688	975.10
45	0.9952	40.5348	965.65
40	0.8593	36.3822	922.04
35	0.707	32.1202	911.12
30	0.6503	30.3954	904.71
25	0.5086	26.3024	889.03
20	0.5424	21.3414	877.32
15	0.4023	18.922	860.76
10	0.3692	15.3898	858.54
05	0.2448	10.0613	853.91

nodes. When the transmission range is varied from the up value to the bottom value, the localization error decreases, respectively, and continues to diminish up to an important value under 0.2 m, also, the convergence accuracy is so remarkable after 20 m, the interpretation of this diminution is due for two grounds—the first one is that more than the spiral anchor node reproach to the other corners beacons nodes more than, the possibility to discover unknowns node grows, and the second one is that each localized nodes at the actual epoch will be well-respected in the upcoming epoch and play as landmark nodes, so will reinforce automatically the localization accuracy.

The value of ALT versus the variation of communication range is presented in the second column of Table 3; we show that the localization time decreases according to the decrease of the communication range, and the phenomena are interpreted that the effectiveness of the spiral node to the settled unknown is very useful; on the other hand, the localized nodes at actual iteration are considered as landmarks nodes at iteration. On the other hand, the algorithm converges rapidly and the energy of the sensors nodes will be consumed so maximized the life time of the network.

Energy consumption (EC) according to the transmission range performs is presented in the third column; we show that when the transmission range is strong, the node requires more power to reach its neighbor's nodes, i.e., a great deal of energy is consumed and vice versa; furthermore, from iteration to another we found that the consumed energy is low; this is due to several reasons:

The route taken by the mobile node allows it not to lose a lot of energy, also, the mobile node strategy processed (synchronize). In fact, the multi-stage localization approach that enables the presence of node placed in the next iteration surprisingly helps to preserve and thus optimize network existence. Finally, the amount of detected nodes increases with every iteration and the localized nodes act as an anchor node in the next iteration.

9 Conclusion and Future Work

We presented a node localization problem based on a parallel location diagram and mobile anchors node via a spiral trajectory. The starting point is the gravity wedge. Aided with four sensor nodes placed in the corners, the mobile node evolves all around searching for positions of neighbor nodes using the Bat algorithm. Exploration is performed through changing the coordinates of Bats by updating the pulsation and loudness of each Bat; these speed up the process of localization by a factor of four. Iterative manner, the mobile beacon node serves to get the best precision of the unknown node coordinates; the obtained result has proved that the modification performed on Bat algorithm makes it very efficient. Furthermore, power consumption has indeed been taken account as an impactful metric, this document discusses node localization as the optimization of multi-criteria and combines also energy consumption, and future progress should highlight WSN's scalability and how such a strategy handles massive underwater systems and incorporate a deep learning strategy to enhance the performance of the IoT environment.

References

1. Adnan MA, Razzaque MA, Ahmed I, Isnin IF (2014) Bio-mimic optimization strategies in wireless sensor networks: a survey. Sensors 141:299–345
2. Bitam S, Zeadally S, Mellouk A (2016) Bio-inspired cybersecurity for wireless sensor networks. IEEE Commun Mag 54(6):68–74
3. Sharma N, Bhatt R (2018) Privacy preservation in WSN for healthcare application. Proc Comput Sci 132:1243–1252. http://doi.org/10.1016/j.procs.2018.05.040
4. Boukerche A, Oliveira H, Nakamura E, Loureiro A (2007) Localization systems for wireless sensor networks. IEEE Wirel Commun 14(6):6–12
5. Akyildiz I, Su W, Sankarasubramaniam Y, Cayirci E (2002) Wireless sensor networks: a survey. Comput Netw 384:393–422
6. Kaviarasan R, Amuthan A (2019) Survey on analysis of meta-heuristic optimization methodologies for node network environment. In: 2019 International Conference on Computer Communication and Informatics (ICCCI), Coimbatore, Tamil Nadu, India, pp 1–4
7. Yang XS (2011) Bat algorithm for multi objective optimization. Int J Bio-Inspired Comput 3(5):267–274
8. Alrajeh NA, Bashir M, Shams B (2013) Localization techniques in wireless sensor networks. Int J Distrib Sens Netw 9(6):1–9
9. Chowdhurya JS, Elkina C, Devabhaktunia D, Rawat DB, Oluochc J (2016) Advances on localization techniques for wireless sensor networks: a survey. Comput Netw 110(9):284–305
10. Miloud M, Abdellatif R, Lorenz P (2019) A novel Bat algorithm for line of sight localization in Internet of Things and wireless sensor network. In: Machine learning and cognitive science applications in cyber security (301118-033546)
11. Zhu J, Chen Q, Zhang J (2017) Localization optimization algorithm of maximum likelihood estimation based on received signal strength. In: 9th IEEE international conference on communication software and networks, 21 Dec 2017, Guangzhou, China, pp 830–834
12. Costa A, Patwari N, Hero AO (2006) Distributed weighted-multidimensional scaling for node localization in sensor networks. ACM Trans Sens Netw 2(1):39–64

13. Gopakumar A, Jacob L (2008) Localization in wireless sensor networks using particle swarm optimization. In: International conference on wireless, mobile and multimedia networks, Beijing, China, pp 227–230
14. Harikrishnan R, Jawahar SK, Sridevi P (2014) Differential evolution approach for localization in wireless sensor networks. In: IEEE international conference on computational intelligence and computing research, India, Sept 2014
15. Goyal S, Pattenh M (2015) Flower pollination algorithm based localization of wireless sensor network. In: 2nd international conference on recent advances in engineering and computational sciences, Chandigarh, India
16. Miloud M, Abdelatif R, Lorenz P (2017) Metaheuristic RSSI based for node localization in distributed wireless sensor network. In: Global Information Infrastructure and Networking Symposium (GIIS), 25–27 Oct 2017, pp 64–70
17. Miloud M, Rahmoun A, Lorenz P, Lasla N (2017) An effective Bat algorithm for node localization in distributed wireless sensor network. Secur Privacy e20. https://doi.org/10.1001/spy2.7
18. Rajakumar R, Amudhavel J, Dhavachelvan P, Vengattaraman T (2017) GWO-LPWSN: grey wolf optimization algorithm for node localization problem in wireless sensor networks. J Comput Netw Commun
19. Strumberger I, Beko M, Tuba M, Minovic M, Bacanin N (2018) Elephant herding optimization algorithm for wireless sensor network localization problem. In: IFIP advances in information and communication technology, pp 175–184
20. Miloud M, Abdellatif R, Lorenz P (2019) Moth flame optimization algorithm range-based for node localization challenge in decentralized wireless sensor network. Int J Distrib Syst Technol (IJDST) 10(1):82–109
21. Lotfi B, Abdellatif R, Lorenz, P, Miloud M (2019) Whale optimization approach for optimization problem in distributed wireless sensor network. In: 9th international conference on information systems and technologies, Article No. 19, Cairo, Egypt, 24–26 Mar 2019. https://doi.org/10.1145/3361570.3361594
22. Qin F, Wei C, Kezhong L (2010) Node Localization with a mobile beacon based on ant colony algorithm in wireless sensor networks. In: International conference on communications and mobile computing, Shenzhen, pp 303–307. https://doi.org/10.1109/cmc.2010.21
23. Philip TD, Asatilla A, Jae-Min L, Dong SK (2019) Mobile node localization using PSO with exploitation of the worst particle. In: KICS Summer conference 2019, Jeju, South Korea
24. Abdi F, Haghighat AT (2014) A hybrid RSSI based localization algorithm for WSN using a mobile anchor node. In: International conference on computing, communication and networking technologies, China. https://doi.org/10.1109/icccnt.2014.6963058
25. Lin Y, Tao H, Tu Y, Liu T (2019) A node self-localization algorithm with a mobile anchor node in underwater acoustic sensor networks. IEEE Access 7:43773–43780
26. Kannadasan K, Edla DR, Kongara MC, Kuppili V (2019) M-curves path planning model for mobile anchor node and localization of sensor nodes using dolphin swarm algorithm. Wirel Netw (May 2019)
27. Shilane D, Martikainen J, Dudoit S, Ovaska S (2008) A general framework for statistical performance comparison of evolutionary computation algorithms. Inf Sci 178:2870–2879

Design of a Virtual Channel Router Architecture for Low Power on Mesh-of-Grid Topology for Network on Chip

K. Somasundaram

Abstract System on chip (SoC) cannot handle the large communication systems due to increasing number of applications. Network on chip (NoC) is an alternative technology to system on chip (SoC) to improve the performance. Still, NoC faces a lot of challenges in its design and testing. The main challenges are power, area, and latency constraints. Router places an important role in communication between the cores. Design of a router is an integral part of a NoC system design. In this paper, we have considered the mesh-of-grid (MoG) topology. We have proposed an architecture for virtual channel router (VCR) to reduce the area and power. Region-based routing (RBR) mechanism is used here to optimize the area and power. We have evaluated various network parameters using the Network Simulator-2. Our experimental results show that the architecture with RBR mechanism gives a significant reduction in area and power in comparison with table-based mechanism and other mechanism. We have shown that the MoG topology gives better latency and packet delivery ratio than the mesh topology.

Keywords Network on chip · Mesh-of-grid topology · Region-based routing · Virtual channel router

---

K. Somasundaram (✉)
Department of Mathematics, Amrita School of Engineering, Amrita Vishwa Vidyapeetham, Coimbatore, India
e-mail: s_sundaram@cb.amrita.edu

S. M. Thampi et al. (eds.), *Applied Soft Computing and Communication Networks*, Lecture Notes in Networks and Systems 125,
https://doi.org/10.1007/978-981-15-3852-0_5

1 Introduction

1.1 Basic Ideas

In recent years, network on chip (NoC) becomes powerful technology for a scalable solution to solve the problems of large-scale communications in multiprocessor system on chip (SoC). It is considered as a subsystem that connects the IP cores in a SoC. NoC brings significant improvements over the conventional bus and crossbar interconnections. SoC consists of memory blocks, dedicated hardware, FPGA blocks, DSP processors, etc., interconnected along with application-specific cores and special circuitry in a complex way. Due to its bus-based architecture, communication bottlenecks are created. Thus, a system with efficient interconnection structure is required.

Design of 3D NoCs is challenging, requiring specialized routing algorithms to enable high performance and prevent deadlocks. Figure 1 illustrates well-known 2D mesh NoC architecture of the size 4×4. As shown in Fig. 1, NoC consists of network interfaces (need for connection between IP cores and topology), cores, routers, and links. The links connect adjacent nodes to each other with either static or dynamic bandwidth. Routing patterns are done through the routers.

To date, several prototype NoCs have been designed in both academia and industry, but only very few have been realized on silicon. Commercial providers on NoC solutions are NetSpeed Systems, Arteris, Sonics, etc. Industry's first network on chip was commercialized by Arteris. Arteris provides network-on-chip interconnect semiconductor processing elements to SoC makers to reduce cycle time. Arteris network-on-chip interconnect technology is being used in Samsung in its mobile phone applications, modems, and processors to easily connect them. This shortens

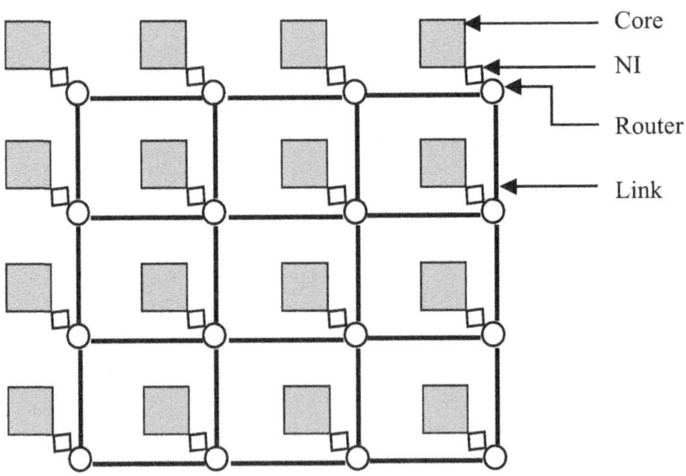

Fig. 1 4×4 2D mesh

the development cycles and reduces IP compatibility risks. SonicsGN (SGN) is an on-chip network allowing advanced SoC communications network design using a high-speed fabric topology structure which is very much scalable. It is Sonics' fourth generation and is configurable. SGN provides NoC with the highest frequency available today. NetSpeed Systems provide coherent, scalable, network-on-chip IPs to SoC designers for applications ranging from mobile to networking and high-performance computing.

1.2 Previous Works

Scalability is an important aspect in NoC topologies. Introductory concepts and recent research works of NoC can be found in the books [1–3]. Exhaustive list of references can be found in some NoC bibliographies, which are available in online. A comprehensive introduction to NoCs and existing design practices are presented in [4]. Application mapping is one of the most important dimensions in NoC research. One can refer a survey on application mapping in [5]. A survey of NoC tools is presented in [6]. Ginosar and Chatha [7] highlighted in their editorial introduction on various issues related to NoC performance, power, and area.

Three-dimensional ICs have several advantages. Merits and demerits of 3D Ics are discussed in [8]. A redundant TSV architecture with reasonable cost for ASICs was proposed by Hsieh et al. [9]. Feero et al. [10] proposed 3D network structures for better performance. Different topologies have been proposed for on-chip interconnection. Deadlock-free routing algorithms for minimizing the data pack transmission in 2D NoC and 3D NoC are given in [11] and [12], respectively. A NoC has many data links connected by routers through point-to- point interconnections. Routers form the back bone of network on chips. Messages can be transmitted from any source core to any destination core over several links, by making switching decisions at the routers. The signals are shared among wires in the links of the NoC. As a result, high level of parallelism is achieved as all links in the NoC can operate parallel on different data packets. In [13], Frank Olaf et al. have shown an efficient and low-cost routing framework for convex mesh partitions. Onizawa et al. [14] proposed a high-throughput and compact asynchronous NoC router which is based on level-encoded dual-rail (LEDR). The LEDR encoding is used for both communication links and routers. In a NoC, limitations in critical path delay give bandwidth issues. Switch port buffer size will cause the critical path delay is heavily. Hence, we believe that by removing the buffers we can increase the switch clock frequency. In [15], Crispín Gómez et al. proposed a technique to reduce packet dropping during the switch process in a bufferless NoC.

Wang and Bagherzadeh [16] proposed a QoS-aware and congestion-aware NoC router architecture that enables quality-oriented network transmission and maintains a feasible implementation cost. Also, it balances traffic load inside the network to enhance overall throughput. In [17], Seitanidis et al. have proposed elastic buffer (EB) architecture to support multiple virtual channels (VCs).

Dimension-order routing (DOR) algorithm will route the packets by comparing the row and the column of the destination switch. Palesi et al. [18] have used a table-based switching approach that stores the output ports for reaching the destination switch. The same design can be used for implementing any routing algorithm and topology. Mejia et al. [19] proposed a region-based routing mechanism for routing packets in 2D structures like meshes with faulty nodes. This consumes very less power and requires lesser area than the conventional method. In this paper, we propose a region-based routing mechanism for storing the output ports corresponding to the destination address and this is applied to a 3D mesh-of-grid topology. We show that the region-based routing mechanism reduces the cost of the table-based switching method.

2 Mesh-of-Grid Topology

An optimized 3D NoC topology is characterized by different trade-offs with regard to network latency, energy consumption, and limited vertical links. The topology should be easily extendable with minimal changes in interconnection links and nodes, in order to adopt the topology for mapping homogeneous and heterogeneous IP blocks of various applications. Vishwanathan et al. [20] showed 3D recursive and 3D modified mesh topologies. The performances of these topologies in terms of buffer size, average flits delay, and dissipated energy were analyzed. Recently, Poovendran and Sumathi [21] showed a low-power SCM topology for high-performance network-on-chip (NoC) architecture using an optimized routing design. They showed that their topology will optimize the area and power.

A mesh-of-grid (MoG) network consists of cores arranged in a grid format. That is, MoG has several meshes arranged in the form of grids. This 3D topology has several advantages in terms of latency and packet loss. In contrast, many communication networks and the protocols are designed to cope with packet loss. For example, the TCP protocol secures the communication against packet loss and thus provides reliable communication. In general, this loss is not caused by unreliable communication media or by buffer overflow.

In a delay model, packets can be delayed but not dropped and whereas the packets can be dropped in a loss model. When a mesh topology is used for sending packets, this takes a longer route to send packets, as a result latency increases. In the case of a MoG, the meshes are arranged one above the other and connected with the help of TSVs. These TSVs make possible the transfer of packets between different layers of topology, and this reduces the distance between the nodes, as a result, the latency reduces. Jitter is defined as a variation in the delay of received packets. Mesh of grid has larger jitter and packet delivery ratio.

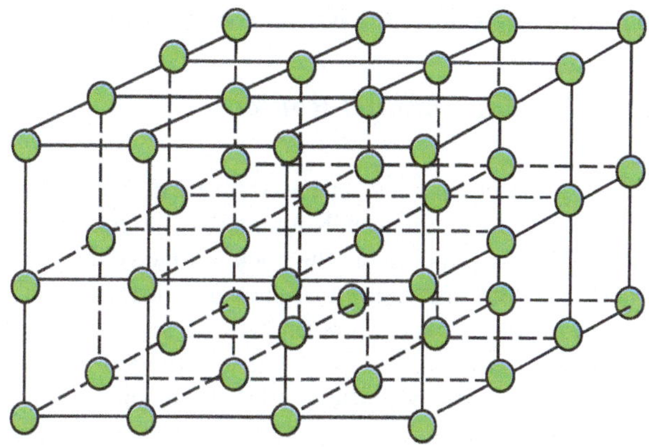

Fig. 2 Three-layer 4 × 4 mesh-of-grid topology

A three-layer 4×4 mesh-of-grid topology is shown in Fig. 2 In a MoG, there are seven ports, namely north port, west port, south port, east port, internal port, and two ports to the adjacent layers. Figure 3 shows the seven ports model. One of the seven output ports will be selected depending on the destination address. Here, 128 bits of 64 words are assumed as input. We design a router with 7 ports.

Three virtual channels associated with each input port. If access is allowed to the crossbar, then the grant signal is sent to the encoder by the round robin arbiter. Depending on the grant signal, the encoder generates an output so as to pass the input to the output port. Encoder assures that one input port can drive only one output port at a time.

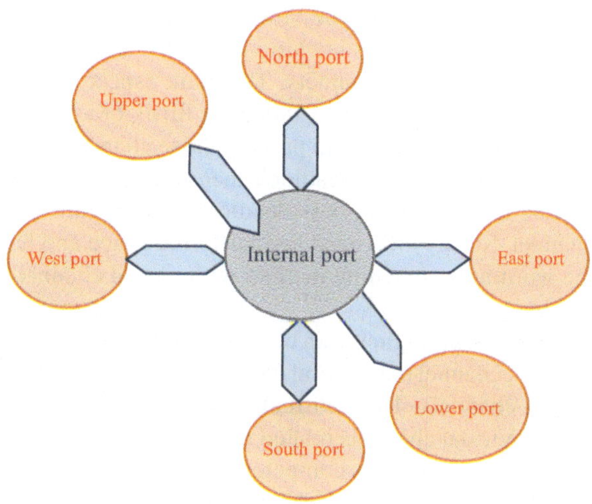

Fig. 3 Seven port router model

3 Virtual Channel Router and Region-Based Routing

3.1 Virtual Channel Router

Router is based on a packet-based protocol whose main function is to route packets from source to destination. Router sends the packet to the output port corresponding to the address contained in the packet. In a routing, every packet split into flits. Flits do not contain any routing or sequence information and have to follow the route for the whole packet. Each flit will be segmented into several small transmitting units, and the message signal will be divided into head flit, body flit, tail flit and then transmit. Shared-buffer router shares the buffer and processes it for the better routing [22]. Alazemi et al. [23] proposed a possible routerless NoC. They explored a new class of on-chip networks, referred to as routerless NoCs, where routers are completely eliminated, that is, to utilize on-chip wiring resources smartly to achieve comparable hop count and scalability as router-based NoCs. Li et al. [24] proposed to schedule packets in input buffers utilizing reorder buffer (RoB) techniques. They design VCs as RoBs to allow packets located not at the head of a VC to be allocated before the head packets.

The routing algorithms could be either deterministic routing or adaptive routing. XY routing is a deterministic routing, which routes between given pair of nodes. Wormhole switching splits the packets into flits for transmission. A disadvantage of wormhole switching is that the number of resources occupied by the packet before reaching its destination will depend on the number of flits in the packet. If the header flit gets blocked by congestion, the whole chains of flits are stalled. This deadlock condition stalls network since all buffers become full and the circular dependency will occur between nodes. The virtual channels are introduced to present deadlock-free routing in wormhole switching networks. This method splits one physical channel into several virtual channels. The main aspect of the virtual channels is the router need not be blocked even though a flit is blocked. The VCR can switch to other non-blocked channels and process those channels until the situation that caused the blocking is gone. By deadlock theorem [25], if a network's dependence graph is acyclic then the network is deadlock free. Figure 4 shows the concept of a virtual channel. Since NoCs need less buffering space and have a low latency requirement, the wormhole switching method with virtual channel is the most suitable switching method.

Network throughput can be enhanced by splitting the buffer storage related to each network channel into several queues. Each physical channel is associated with several virtual channels. These virtual channels are small queues rather than a single deep queue. Virtual channels related to one physical channel are allocated independently. These virtual channels compete with each other for physical bandwidth. It separates transmission resources and buffer resources. These decoupling by-passes blocked messages so that active messages can pass using channels that would otherwise be left idle. When the head flit arrives, it arbitrates with other virtual channels on the particular input port. If it wins, it will be selected for forwarding to the output port.

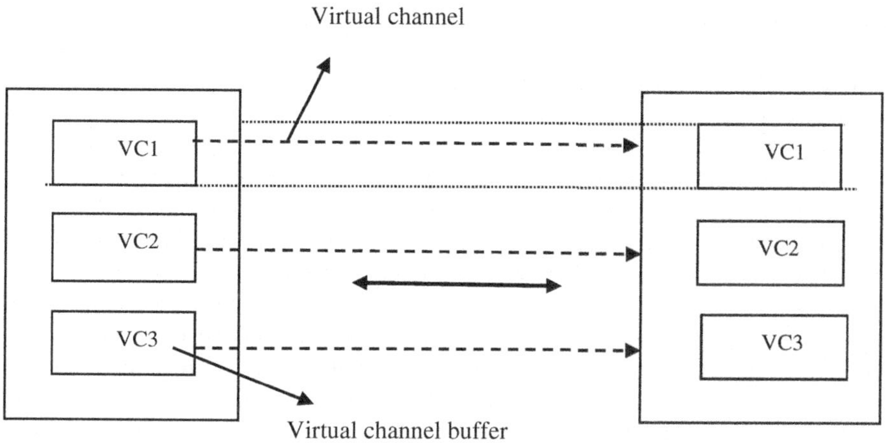

Fig. 4 Virtual channel router

For each input port connection, the output port has a single stage of buffering. There will be arbitration among the flits in the buffer to access the link to the input controller on the next core. Region-based routing mechanism is used to find the port address depending on the destination address.

Young Jin et al. [26] have shown a comparative analysis of NoC implementations based on virtual channels (VC) versus multiple physical (MP) networks. They explored the design space considering area occupation, target clock period, and power dissipation, as well as system-level performance metrics such as average latency and maximum sustained throughput. Pournamy Mohan and Somasundaram [27] shown a router architecture, which is based on quality of service (QoS) mechanism. They considered a partial mesh-of-grid topology, and the router was proposed to minimize the area and power. In [28], Somasundaram and Chaithanya proposed a routing algorithm for minimizing the congestion factor during the routing. Also, they proposed a router for a topology which will minimize the area and power. Yazdanpanah et al. [29] proposed an energy-efficient partition-based XYZ-planar routing algorithm for a wireless network on chip.

3.2 *Region-Based Routing*

Adaptive routing is one of the well known routing methods in NoC. In recent years, different fully or partially adaptive routing algorithms were proposed by several authors. The adaptive routing focuses only on local traffic. Mak et al. [30] gave path planning techniques for adaptations. In [31], the authors were proposed universal logic-based distributed routing (uLBDR)) for irregular topologies. There are some limitations on LBDR. To overcome the limitations of LBDR, we propose a region-based routing mechanism.

In region-based routing, routing is done by grouping the destinations into regions. This is an efficient method to design cost-effective router. The router consists of input and output ports, crossbar, and an arbiter. Different blocks are required for wormhole switches. The input buffer will be managed by input port controller. The packet header should be as compact as possible in order to keep the minimum overhead. Along with other informations, the packet header contains the destination identifier of the packet. In this, absolute addressing is done and the address given in the packet will identify the *X*-coordinate and *Y*-coordinate of the destination core. These are represented as row destination (RowDst) and column destination (ColDst), respectively. The destination coordinates are identified and compared with the coordinates of the current switch. If it is equal, packet is delivered to the internal node, otherwise the coordinates are sent to the routing control unit in the hardware.

The routing control (RC) unit has three sub-modules: the region matcher, region trigger, and output port selector. For example, consider a two-layer 4 × 4 MoG with each router having six input ports and six output ports. As it has only one adjacent layer, only one port is used for transferring packets to adjacent layer. Figure 5 shows how the regions are defined. Here, we take the ninth node as the node under consideration. Every node which belongs to the north side of the node 9 can be reached through north port. So, these areas can be grouped onto a region. This forms the Routing Region1 (RR1). Nodes that can be accessed through the east output port is grouped into a single region which forms the Routing Region2 (RR2). The entire nodes in the next layer are grouped into a region. Similarly, all regions are defined according to the ports that have to be taken to reach the destination node. Now, there may be some nodes that can be accessed using more than one port. In order to avoid confusion, we treat them as if they can be reached only through one port. For instance, take five regions, each routing region is defined using the coordinates of top leftmost nodes and bottom rightmost nodes. Col1RR1 and Row1RR1 are used to define the

Fig. 5 Defining regions

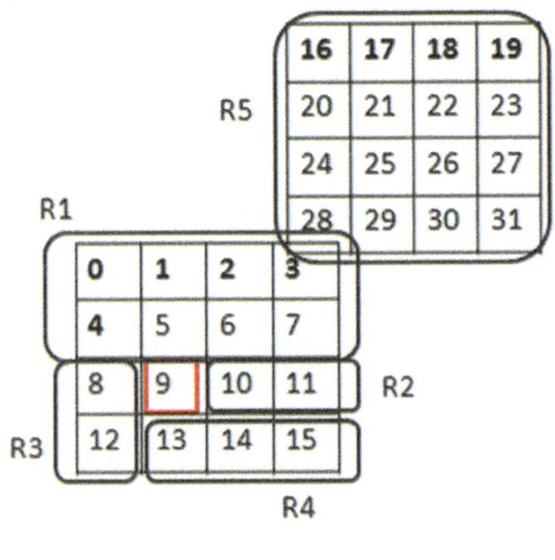

coordinates of top leftmost node and Col2RR1 and Row2RR1 are used to define the coordinates of bottom rightmost node. LayerRR1 is used to define the layer of routing region RR1. Similarly, the coordinates of all routing regions are defined. The pseudocode for region-based routing (RBR) mechanism is shown in the Algorithm 1. Here, the pseudocode is defined for a generic MoG topology with seven input ports and seven output ports as it has seven routing regions. In the pseudocode, the coordinates of the destination nodes are compared with the coordinates of the routing region. These comparisons are done using RC unit. If the conditions are satisfied, it goes to the inner loop and sets a destination port. This is checked for every port as the value of variable N changes from 0 to $n - 1$ where n is the number of ports. Depending on the routing region to which the destination node of the packet in an input port belongs to, an output port will be selected for each input port. There are seven ports, north (N), south (S), east (E), west (W), internal (I), down (D), and up (U). We denote *src* as a source node and *dst* as a destination node.

We are interested to minimize power. In this process, we select the regions corresponding to input port registers that do not match with the input port of the packet being routed and remove the regions. Region trigger (RT) unit does this function. RT correlates the input port register of every region to the input port of the packet. The RT unit consists of six AND gates and a single OR gate. Figure 6 shows the proposed hardware implementation.

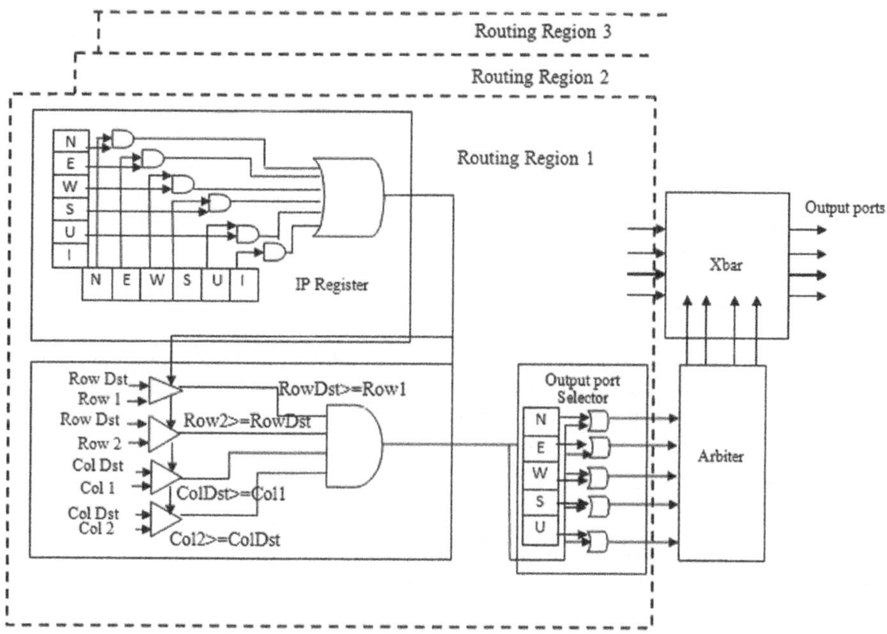

Fig. 6 Hardware implementation of RBR mechanism

Algorithm 1: The Pseudo code of RBR mechanism

```
For N varies from 0 to n-1
  {
    for src, dst ∈ Topology, src ≠ dst
    {
      {
        if Routing_Region_Trigger_1
       if (dstLayerN<=layerRR1) & (RowDstN>=Row1RR1) & (RowDstN<=Row2RR1)
         &(ColDstN>=Col1RR1) & (ColDstN<=Col2RR1)
       then set destp_ip_portN as N
      }
     if Routing_Region_Trigger_2
      {
        if (dstLayerN<=layerRR2) & (RowDstN>=Row1RR2) & (RowDstN<=Row2RR2) &
        (ColDstN>=Col1RR2) & (ColDstN<=Col2RR2)
         then set destp_ip_portN as E
      }
     if Routing_Region_Trigger_3
      {
        if (dstLayerN<=layerRR3) & (RowDstN>=Row1RR3) & (RowDstN<=Row2RR3) &
        (ColDstN>=Col1RR3) & (ColDstN<=Col2RR3)
       then set destp_ip_portN as W
      }
     if Routing_Region_Trigger_4
      {
        if (dstLayerN<=layerRR4) & (RowDstN>=Row1RR4) & (RowDstN<=Row2RR4) &
        (ColDstN>=Col1RR4) & (ColDstN<=Col2RR4))
        then set destp_ip_portN as S
      }
     if Routing_Region_Trigger_5
      {
        if (dstLayerN<=layerRR5) & (RowDstN>=Row1RR5) & (RowDstN<=Row2RR5) &
        (ColDstN>=Col1RR5) & (ColDstN<=Col2RR5)
         then set destp_ip_portN as U
      }
    if Routing_Region_Trigger_6
      {
        if (dstLayerN<=layerRR6) & (RowDstN>=Row1RR6) & (RowDstN<=Row2RR6) &
        (ColDstN>=Col1RR6) & (ColDstN<=Col2RR6))
        then set destp_ip_portN as D
       }
    }
 for src, dst ∈ Topology, src = dst
    {
      if Routing_Region_Trigger_5
     if dstLayerN>=layerRR5 then destp_ip_portN = I
    }
  }
End.
```

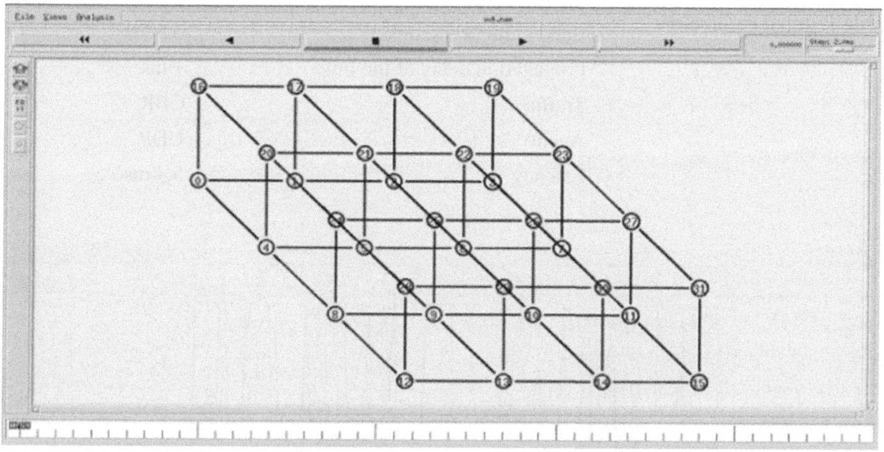

Fig. 7 Snapshot of the NS-2 network animator showing a MGT

The RT unit for the outputs gives information for the remaining logic for the region. For checking the regions corresponding to destination core, we use four comparators. All the regions are evaluated in parallel and the output ports are ORed and passed to the arbiter. Arbiter arbitrates one port after the other. Thus, the packet is passed until it reaches the destination. A table with large number of entries will be minimized to a very small number of entries. Each topology gives the requirements of the number of regions and its routing protocols (Fig. 7).

4 Experimental Results

The mesh-of-grid topology is simulated using Network Simulator tool. NS-2 is an object-oriented, discrete event-driven network simulator implemented in C++ and TCL Scripts. Network Simulator tool provides a convenient user interface.

An adaptive routing algorithm is used with RBR in our experiments. The parameters chosen for simulation are (i) packet size of 5 flits with each flit being 64 bits, (ii) input buffer size 12 flits (iii) XY routing and wormhole flow control (iv) simulation cycles count: 6000 and (vi) warm up cycles count: 1000. In our experimental study, we assume the following: injects flits: 20–100 Gbps, switch service rate 400 Gbps, maximum burst size of the data $b = 100$ bytes and flit size $k = 10$ bytes (Table 1).

Bandwidth graph obtained using X-graph is shown in Fig. 8. We take the throughput as Throughput = {(total packets completed) * (packet length)}/{(no. of IP cores) * (total time)}, where the total packets completed refer to the number of packets that successfully arrived at the destination IPs, packet length refers to the number of flits, number of IP cores refers to the number of IP blocks involved in communication, and the total time refers to the total simulation time (in clock cycles).

Table 1 Input parameters

Bandwidth of the link	0.5 Mb
Propagation delay of the link	1 ms
Traffic	CBR
Agent	UDP
Latency	5.4 ms

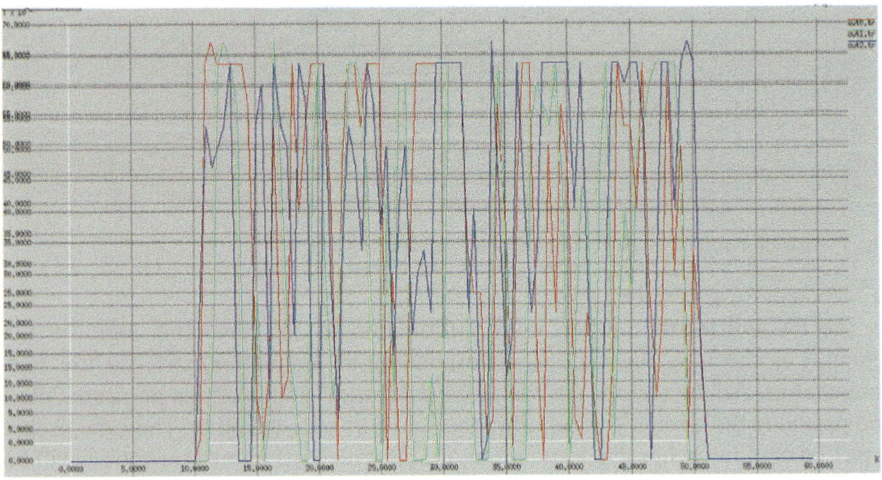

Fig. 8 Bandwidth calculation using *X*-graph

Bandwidth is given by Bandwidth = (throughput * number of cores * flit size)/clock period.

The benefits of VCs plus RBR routing algorithm are (i) we can avoid the packet losses since the VCs will re-route the packets if they block in a particular node. Hence, the packet delivery ratio is very high. (ii) Due to the loopback RBR mechanism, the area and power are minimum compared to table-based switching method. Using NS-2 tool, we obtain the packet delivery ratio as 0.9987. The results are tabulated and compared with the mesh topology. Table 2 shows a comparison between mesh-of-grid

Table 2 NS-2 synthesis report

	Mesh-of-grid topology	Mesh topology
Packet loss	0	6
Latency	0.324 s	0.486 s
Jitter	Send: 802, received: 801	Send: 802, received: 783
Packet delivery ratio	0.9987	0.9763

Table 3 Device utilization summary of MoG topology and mesh topology

Logic utilization	Mesh-of-grid topology	Mesh topology				
	Available	Used	Utilization (%)	Available	Used	Utilization (%)
Number of occupied slices	4656	14	1	4896	5	1
Number of slices containing only related logic	14	14	100	12	12	100
No. of slices containing unrelated logic	14	0	0	12	0	0
Total number of four input LUTs	9312	21	1	4896	17	1
Number of bonded IOBs	190	13	6	172	13	7

topology and mesh topology. From Table 2, we observed that there is an improvement in latency and packet delivery ratio using MoG than with mesh topology. There is a significant reduction in the latency for MoG, and this is due to TSV link. Using Verilog HDL, we implement the MoG-based network and our synthesis is based on Xilinx ISE. Summary of the utilities of the two topologies are shown in Table 3. It is observed that the area occupied is less for MoG network than for mesh network of the same size.

We synthesize the region-based virtual channel router (VCR) in 90 nm technology and we implement this using Verilog HDL. A thorough performance evaluation and estimation of area overhead of VCR has been carried out. Area and power analysis are done using Synopsys Design Vision supporting 90 nm CMOS technology. Table 4 shows a comparison of table-based VCR with region-based VCR. Using Synopsys Design Vision, we obtain the area of table-based VCR as 251,245.3 μm^2 and with RBR mechanism as 182,652.3 μm^2. Other parameters are shown in Table 4.

We observed that the area and power are less for VCR with RBR mechanism than with table-based routing. Table 5 shows the comparison of table-based router with region-based router. The area of router obtained using Synopsys Design Vision with table-based VCR is 88,704.6 μm^2 and for the router with RBR mechanism is 87,424.5 μm^2. Similarly, we compute the other parameters and the values are shown in Table 5. Here also, area and power are less for router with region-based routing

Table 4 Comparisons of table-based VCR with region-based VCR

Cost metrics	Table-based VCR	VCR with RBR
Area	251,245.3 μm^2	182,652.3 μm^2
Cell internal power	6.1871 mW (83%)	1.5636 mW (47%)
Net switching power	1.2719 mW (17%)	1.7613 mW (53%)
Total dynamic power	7.4589 mW (100%)	3.3248 mW (100%)
Cell leakage power	923.6841 μW	600.7408 μW

Table 5 Comparisons of table-based router with region-based router

Cost metrics	Router with table-based VCR	Router with RBR
Area	88,074.6 μm^2	87,424.5 μm^2
Cell internal power	2.6114 mW (94%)	2.3959 mW (94%)
Net switching power	172.3863 μW (6%)	161.0400 μW (6%)
Total dynamic power	2.7838 mW (100%)	2.5570 mW (100%)
Cell leakage power	346.8311 μW	340.4551 μW

than with table-based routing. The GTK waveform for virtual channel router is given in Fig. 9. Reliability Aware Virtual Channel (RAVC) NoC router micro-architecture is shown in [32]. Also, we observed that our RBR outperform the RAVC. In [17], Mejia et al. proposed RBR mechanism in 2D networks like, 4×4, 8×8, and 16×16. Here, we considered 3D network topology along with RBR loopback mechanism. Due to this, area and power are reduced.

Also, we synthesized the area and power with different flit sizes of the proposed router. We use the Synopsys Design Vision. We evaluate the area and power different designs, 128b MUX, 64b MUX, and 32b MUX, which are given in Table 6. Average end-to-end delay for MoG with different data injection rates are shown in Table 7.

5 Conclusions

In this paper, we proposed a RBR-based virtual channel router for minimal power and area. Using Design Vision supporting 90 nm CMOS technology, we synthesis the area and power. The deadlock is avoided by using virtual channel routers. Our experimental results show that the region-based routing consumes lesser area and has

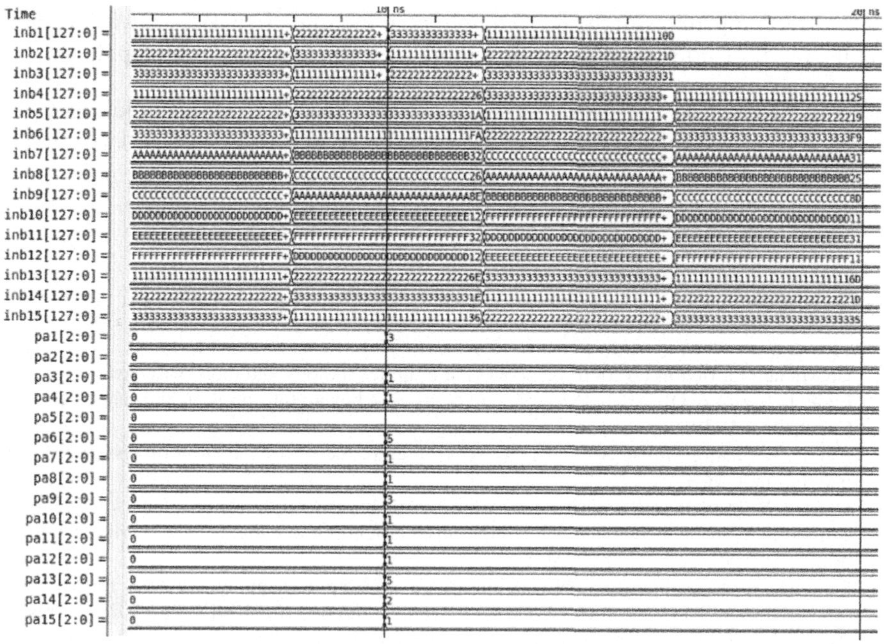

Fig. 9 GTK waveform of virtual channel router

Table 6 Comparison of area and power of router with different flit sizes

Flit size	Area	Power (mW)
32b MUX	67,490	1.863
64b MUX	115,010	3.222
128b MUX	211,704	5.778

Table 7 Average end-to-end delay of MoG with various data injection rates

Injection rate in Gbps	Average end-to-end switch delay in ps	Average link delay in ps	Average end-to-end delay in ps
20	19,799.8	500	20,299.8
40	20,083.6	500	20,583.6
60	20,365.36	500	20,865.36
80	20,648.26	500	21,148.26
100	20,930.5	500	21,430.5

lower power requirement compared with the table-based routing. Also, we prove that the mesh-of-grid topology has a better latency, larger packet delivery ratio, and less packet loss compared with the mesh topology of the same size. Hence, we conclude that the mesh-of-grid-based network is more area efficient than mesh topology.

Acknowledgements Author gives his sincere thanks to Ms. Anjali N.V. (M.Tech VLSI student) for her help in experimental results. The author also thanks the anonymous reviewers for their valuable suggestions.

References

1. De Micheli G, Benini L (2006) Networks-on-chip: technology and tools. Elsevier
2. Flich J, Bertozzi D (2010) Designing network-on-chip architectures in the nanoscale era. Chapman & Hall/CRC
3. Kundu S, Chattopadhyay S (2018) Network-on-chip: the next generation of system-on-chip integration. CRC Press
4. Ogras UY, Marculescu R (2013) Modeling, analysis and optimization of network-on-chip communication architectures. Lecture notes in electrical engineering, vol 184. Springer. https://doi.org/10.1007/978-94-007-3958-i
5. Sahu PK, Chattopadhyay S (2013) A survey on application mapping strategies for network-on-chip design. J Syst Architect 59(1):60–76. https://doi.org/10.1016/j.sysarc.2012.10.004
6. Achballah AB, Saoud SB (2013) A survey of network-on-chip tools. Int J Adv Comput Sci Appl 4(9):61–67. https://arxiv.org/ftp/arxiv/papers/1312/1312.2976.pdf
7. Ginosar R, Chatha KS (2014) Guest editors' introduction—special issue on network-on-chip. IEEE Trans Comput 63(3):527–528. https://doi.org/10.1109/mdt.2005.136
8. Davis WR, Wilson J, Mick S, Xu J, Hua H, Mineo C, Sule AM, Steer M, Franzon PD (2005) Demystifying 3D ICs: the pros and cons of going vertical. IEEE Des Test Comput 22(6):498–510
9. Hsieh AC, Hwang TT, Chang MT, Tsai MH, Tseng CM, Li HC (2010) TSV redundancy: architecture and design issues in 3D. In: Proceeding of the conference Design Automation and Test in Europe (DATE), pp 166–171. https://doi.org/10.1109/tvlsi.2011.2107924
10. Feero BS, Pande PP (2009) Networks-on-chip in a three-dimensional environment: a performance evaluation. IEEE Trans Comput 58(1):32–45. https://doi.org/10.1109/TC.2008.142
11. Somasundaram K, Plosila J (2012) Deadlock free routing algorithm for minimizing data packet transmission in network on chip. Int J Embedded Real Time Commun Syst 3(1):70–81. https://doi.org/10.4018/jertcs.2012010104
12. Somasundaram K, Plosila J, Vishwanathan N (2014) Deadlock free routing algorithm for minimizing the congestion in a hamiltonian connected recursive 3D-NoCs. Microelectron J 45(8):989–1000. https://doi.org/10.1016/j.mejo.2014.05.003
13. Sem-Jacobsen FO, Rodrigo S, Skeie T, Strano A, Bertozzi D (2013) An efficient, low-cost routing framework for convex mesh partitions to support virtualization. ACM Trans Embedded Comput Syst 12(4):1–24. https://doi.org/10.1145/2485984.2485995
14. Onizawa N, Matsumoto A, Funazaki T, Hanyu T (2014) High-throughput compact delay-Insensitive asynchronous NoC router. IEEE Trans Comput 63(3):637–649. https://doi.org/10.1109/TC.2013.81
15. Gómez Requena C, Gómez ME, López P, Duato José (2011) How to reduce packet dropping in a bufferless NoC. Concurrency Comput Prac Experience 23:86–99. https://doi.org/10.1002/cpe.1606
16. Wang C, Bagherzadeh N (2014) Design and evaluation of a high throughput QoS-aware and congestion-aware router architecture for network-on-Chip. Microprocess Microsyst 38(4):304–315. https://doi.org/10.1016/j.micpro.2013.09.006

17. Seitanidis I, Psarras A, Dimitrakopoulos G, Nicopoulos C (2014) ElastiStore: an elastic buffer architecture for network-on-chip routers. In: Proceeding of DATE14. https://doi.org/10.7873/date.2014.253
18. Palesi M, Kumar S, Holsmarg R (2006) A method for router table compression for application specific routing in mesh topology NoC architectures. In: Proceedings of SAMOS, pp 373–384. https://doi.org/10.1007/11796435_38
19. Mejia A, Palesi M, Flich J, Kumar S, López P, Holsmark R, Duato J (2009) Region-based routing: a mechanism to support efficient routing algorithms in NoCs. IEEE Trans Very Large Scale Integr (VLSI) Syst 17(3):356–369. https://doi.org/10.1109/tvlsi.2008.2012010
20. Vishwanathan N, Paramasira R, Somasundaram K (2014) An optimized 3D topology for on-chip communications. Int J Parallel Emergent Distrib Syst 25(4):346–362. https://doi.org/10.1080/17445760.2013.866236
21. Poovendran R, Sumathi S (2019) An area-efficient low-power SCM topology for high performance network-on Chip (NoC) architecture using an optimized routing design. Concurrency Comput Pract Experience 31(14)
22. Yoon YJ, Concer N, Petracca M, Carloni LP (2013) Virtual channels and multiple physical networks: two alternatives to improve NoC performance. IEEE Trans Comput Aided Des Integr Circ Syst 32(12):1906–1919. https://doi.org/10.1109/TCAD.2013.2276399
23. Alazemi F, AziziMazreah A, Bose B, Chen L (2018) Routerless network-on-chip. In: 2018 IEEE international symposium on High Performance Computer Architecture (HPCA). IEEE, pp 492–503
24. Li C, Dong D, Z Lu, Liao X (2018) RoB-router: a reorder buffer enabled low latency network-on-chip router. IEEE Trans Parallel Distrib Syst 29(9):2090–2104
25. Verbeek F, Schmaltz J (2014) A decision procedure for deadlock-free routing in wormhole networks. IEEE Trans Parallel Distrib 25(8):1935–1944. https://doi.org/10.1109/TPDS.2013.121
26. Ben-Itzhak Y, Cidon I, Kolodny A, Shabun M, Shmuel N (2014) Heterogeneous NoC router architecture. IEEE Trans Parallel Distrib Syst 26(9):2479–2492. https://doi.org/10.1109/TPDS.2014.2351816
27. Mohan P, Somasundaram K (2015) Design and evaluation of 3D NoC routers with quality-of-service (QoS) mechanism for multi-core systems. In: Proceedings of ICAIECES 2015, Artificial Intelligence and Evolutionary Computations in Engineering Systems, pp 429–441. https://doi.org/10.1007/978-81-322-2656-7_39
28. Kanagasabapathi S, Chythanya C (2019) A routing algorithm and a router architecture for 3D NoC. Comput Sci 20(3)
29. Yazdanpanah F, AfsharMazayejani R, Alaei M, Rezaei A, Daneshtalab M (2019) An energy-efficient partition-based XYZ-planar routing algorithm for a wireless network-on-chip. J Supercomputing 75(2):837–861
30. Mak T, Cheung PYK, Lam K-P, Luk W (2011) Adaptive routing in network-on-chips using a dynamic-programming network. IEEE Trans Industr Electron 58(8):3701–3716
31. Rodrigo S, Flich J, Roca A, Medardoni S, Bertozzi D, Camacho J, Silla F, Duato J (2010) Addressing manufacturing challenges with cost-efficient fault tolerant routing. In: Proceeding of NOCS-2010, pp 25–32. https://doi.org/10.1109/nocs.2010.12
32. Neishaburi MH, Zilic Z (2009) Reliability aware NoC router architecture using input channel buffer sharing. In: Proceeding of GLSVLSI'09, pp 511–516. https://doi.org/10.1145/1531542.1531658

Agent-Based Traffic Obstacles Information System

Kamil Ruta, Damian Rakus, Maria Ganzha, and Marcin Paprzycki

Abstract Traffic management is one of the important aspects of modern cities. With the advent of so-called *Smart Cities*, it is claimed that information technologies can be used to deliver a substantial improvement in vehicle flow. As a result, among others, travel time and CO_2 generation can be reduced. In this paper, we present an agent-based traffic obstacles information and avoidance system. We discuss the advantages of such a system, its key components, and technologies used during implementation. We also present sample scenarios, illustrating how the implemented system works.

Keywords Agents · Traffic · Traffic jam · Agent-based · Agent system · JADE · Java

1 Introduction

Nowadays, one of the important issues is: how to deal with urban traffic. This question concerns all large cities around the world regardless of their location, wealth, or ethnicity. There are at least two main issues related to urban traffic. First, travel time; here, traveling in a car may take more time than traveling, the same road, using a bicycle. Second, pollution; hundreds (sometimes even tens of thousands) of cars are barely moving due to traffic jams, generating enormous amounts of CO_2 (and other pollutants) that remain trapped within the city.

It is believed that information technologies, realizing the concept of *Smart City* [1] can, among others, improve traffic management. This claim has been partially validated by "long-distance travel" applications, such as in-car navigation connected to traffic warning systems, or Google maps. However, one of their disadvantages is the speed of reaction to changing road conditions. Quite often, warning about (new)

K. Ruta · D. Rakus · M. Ganzha · M. Paprzycki (✉)
Warsaw University of Technology, Warsaw, Poland
e-mail: marcin.paprzycki@ibspan.waw.pl

M. Ganzha · M. Paprzycki
Systems Research Institute, Polish Academy of Sciences, Warsaw, Poland

S. M. Thampi et al. (eds.), *Applied Soft Computing and Communication Networks*, Lecture Notes in Networks and Systems 125,
https://doi.org/10.1007/978-981-15-3852-0_6

traffic jam, due to an accident, is "delayed," or warning about non-existent disruption remains. Such situations have a lesser impact when considering long-distance highway travel. Here, traffic information can be propagated across longer distances, while cars between two highway exits have "no way to escape." The situation is much more complex in urban traffic, where a large number of crossing roads, and a very large number of traveling cars make a difference. On the one hand, it is much more difficult to predict where each given car will move at the next crossing. On the other, large number of possible ways of traveling from "point A to point B" add potential flexibility to any traffic management system.

Keeping this in mind, let us consider *software agents*. As can be seen in Sect. 2, they are very well suited when transportation schedule is negotiated and, in case of disruptions, renegotiated. The latter case directly deals with the situation when a car, moving toward a specific location, has to adapt its route due to the dynamically changing road conditions. Agents' ability to rapidly exchange information is a significant factor because it may prevent situations when user is informed about road difficulty too late and has to go back in order to bypass the obstacle.

Therefore, we have decided to investigate how software agents can be used for traffic management in a Smart City. Here, for instance, when a car detects an accident, an agent representing it reports this fact to the "central agent," which informs other interested car agents about the obstacle. By `obstacle`, we mean any "serious difficulty" on the route, such as traffic accident or traffic jam. Next, the state of the "obstacle" is to be regularly checked and updates are issued if situation changes. In the proposed system, when driver specifies the destination, car agent obtains information about potential obstacles, to try to avoid them. Moreover, cars can obtain "instant information" about "new problems," so their route can be dynamically adjusted.

2 State-of-the-Art in Pertinent Areas

Let us now briefly summarize the related state-of-the-art, starting from "driver support systems." The most advanced and popular, current solutions, are Google Maps [2, 3] and Waze [4]. Google Maps allows users to specify one or more destination points and shows the best route (and, possibly, few alternates) to reach the destination. Expected travel time is computed considering current and predicted traffic conditions. Here, Google Maps accesses users' location (calculated using GPS data and cell tower triangulation). Obstacle information is gathered from local offices and Waze databases. Note that travel time prediction is based on current and historical data. Hence, the quality of prediction depends on the "freshness of data." When a street is being closed, or an obstacle appears on the road, there is some time delay before this fact will be "noticed" and considered in calculations. Traffic congestion calculation requires correct input data. Hence, the system works best when large number of users travel with mobile data turned on. Note that "uneven distribution of users," in the city, may affect the precision of predictions delivered by the system.

Waze is a navigation app with data provided by the user's community called Wazers. They report traffic events such as accidents, jams, and street protests. Users can also update more persistent map data, like landmarks, or new roads. Collected data, and anonymous user information, including location and speed, are stored on a central server. Using this information, Waze allows user to find the fastest route for the destination. In 2013, Waze was bought by Google and since then it also shares data with Google Maps.

Apple Maps [5] works in almost exactly the same way. Traffic events are reported mostly by users that are willing to share information and warn others about difficulties on the road. However, in addition to crowdsourced information, Apple is also gathering data from local databases, containing information about traffic accidents and road works. Moreover, Apple Maps track users' average speed based on their GPS location and can determine high traffic congestion.

Let us consider information about obstacles, available in described solutions. Use of Google Maps, or Apple Maps, is associated with storing user-location data. Here, we will immediately note that as will be seen, only data about coordinates of obstacles is stored in the proposed system. The platform also "links the user to the streets" that are a part of the route, but after crossing a given street, this information is deleted. In addition, each city may provide its own platform instance, so none of the data is being stored globally and, in this way, the proposed system can be scaled.

One more solution, similar to Waze, is Yanosik [6], which also provides users with navigation based on, user-generated, almost-real-time traffic data. Specifically, during travel with Yanosik, user can alert other drivers about accidents, road works, police controls, and speed cameras. Here, note that community-driven apps, need some initial number of users to provide useful service for drivers. Moreover, information reported by individual persons may be of "different quality," so the more reports about an issue, the greater certainty that the data is reliable.

Another traffic navigation service is provided by INRIX. Company's app, called 'INRIX Traffic' [7, 8], collects anonymized data on congestion, traffic incidents, parking, and weather-related road conditions, from large number of data points, daily in over 80 countries. Data is aggregated from connected cars, mobile devices, state's Department of Transportation, cameras and sensors on roadways, and major events expected to affect traffic. In addition, INRIX works with local authorities to digitize rules of the road for highly automated vehicles (HAVs) operating on public roads. Moreover, information gathered from HAVs can be used for infrastructure improvements. INRIX's software updates (every 60 s) information about traffic conditions, such as accidents, road works, and speed of traffic in different lanes. It then suggests the fastest route, taking into account all available factors. However, to achieve its wide functionality, a lot of data is being sent and processed. Setting up and maintaining infrastructure consisting of millions of cameras and sensors generates enormous costs and requires permission from various authorities.

Finally, the work presented in [9] proposes another solution to traffic information distribution. Here, information is gathered in key points in a city district. Proposed system is composed of cameras, GPS locators, motion detectors, variable message road signs, and mobile applications. Moreover, detectors are placed at road crossings and help to measure traffic congestion. Available cameras can be used to detect accidents and other obstacles. Information is provided to drivers by road signs and the application. Thanks to the detectors and cameras, it is possible to detect traffic issues almost instantly. This allows to calculate the route, taking into consideration most recent road conditions. However, because of the costs and need of manual work, the system assumes installing cameras only in few "strategic" points. That may lead to incomplete data about "small roads." Here, our proposal is to use drones, so it is not necessary to install extra cameras.

As we can see, current solutions accommodate information about traffic obstacles and effects of congestion. While our system can be extended with this functionality, it was deemed to be outside of scope of our preliminary investigation.

Next, let us summarize key developments in the area of application of software agents in logistics (travel scheduling). In [10], a MAGENTA multi-agent logistic system supporting fleet scheduling is described. The technology is composed of *Ontology Management Toolkit* and *Virtual Market Engine*. The *Ontology Management Toolkit* works as a business knowledge base for agents, representing concepts and interrelationships between them. Example concepts are "Customer," "Cargo" or "CarrierCompany." The *Virtual Market Engine* allows to run agents, and monitor what they are doing. In the studied case, the system was deployed to optimize shipping schedules, as events (e.g., new cargo) occur. As a response to an event, agents start to negotiate an alternative solution, which must satisfy the specified criteria. Providing different criteria for optimization, human scheduler is able to get various reports. Because decisions involve negotiations, human schedulers get the comprehensible rationale for the proposal. Although the schedule is determined based on multiple variables, the system finds an approximation of the "ideal" solution. Here, the use of semantic technologies, while interesting, seems like an overkill, in terms of system complexity and usability, outside of dedicated systems for large fleets.

Next example of transport optimization, with the use of software agents, is presented in [11]. The Whitestein Technologies' system calculates the optimal routes for trucks, while minimizing the cost. Majority of processing, allowing handling almost-real-time changes, is done automatically. Only in special situations, human involvement is needed. The solution may be implemented using agent-based approach in two different ways. Both of them concern the exchange of orders through concurrent negotiations between agents. The first approach is to assign agents to trucks. Then, high granularity and scalability are achieved, but we increase computation time and resource usage, because of a large number of exchanged messages. The second solution is to assign agents to regions, which manage groups of trucks. Here, optimization is performed within regions (first) and between regions (next). This approach produces fewer messages, however, division into regions decreases overall system quality, because initial considerations concerning a new order will take into account only trucks from the order's region. Overall, the agent-based approach has been found to be effective for large fleet transport optimization (e.g., DHL Europe).

On the basis of the above examples, it can be concluded that agent-based approach is worthy of trying in the context of smart city traffic management. Agents reflect the distributed nature of the problem with large number of users, allow the creation of a scalable solution, and provide easy way to exchange information.

3 System Overview

Let us now outline the key aspects of the proposed approach. Let us start from the main assumptions that underline system design. First, we assume that all vehicles are connected to the Internet. This assumption can be questioned. It cannot be expected that anytime soon, all vehicles will be Internet-enabled. However, it is obvious that already today, almost all vehicles, when moving, have on board at least one Internet-connected device (driver's cell phone). Therefore, this assumption is not too far fetched. Second, there exists an infrastructure that allows car devices to communicate with each other. Here, in the near future, 5G networks are to facilitate car-to-car communication [12]. However, let us stress that the way that communication is to be realized is inconsequential for presented work. Finally, we assume that cars will be able to recognize and report various road obstacles. This involves both "human reporting" (now) and "car reporting" (future) capabilities.

Taking into account these general assumptions, let us now outline the design of the proposed `Traffic obstacles information system`. Here, let us note that the *user*, can be either a person driving a car or a system responsible for managing a self-driving car. With this in mind, we have formulated the following requirements for the system:

- User should receive "almost-real-time" information about problems on the road that the vehicle is to travel.
- User will join and listen for information about obstacles on Internet channels representing roads that are to be traversed.
- User should be notified asynchronously without explicitly requesting the data (data push model) only about obstacles that are pertinent to the current route.
- User should receive alternative route if the current is "blocked" by an obstacle.
- Users should be able to inform system about difficulties spotted on the road.

Now, let us define actions that the proposed system should be able to perform:

- Receive information about difficulties on the road (from drivers and cars with installed obstacle detection systems).
- Use drone to check and update the status of the obstacle.
- Report coordinates of all obstacles within specified area (e.g., district, city) on route that user requested.
- Propagate information about recently spotted obstacles to users on Internet channels (e.g., using JADE's Topic) named after streets.

Based on these requirements we can, in the next Section, define technologies that we have decided to use to realize the proposed architecture.

4 Technologies Used

Based on comprehensive analysis of available agent platforms, we have decided to use to JAVA Agent Development Framework (JADE [13]). The system could be created using Spade [14] (multi-agent systems platform written in Python), Jason [15] (an interpreter for an extended version of AgentSpeak) or other similar platforms. However, we decided to use JADE because of its popularity and integration with Java. Currently, the system is not a big programming project, but it may be easily extended to include other city traffic services. Here, managing a static-typed code repository should increase the system stability (e.g., comparing to Spade).

The developed system emulator has been written using Java 8. System GUI has been created using JavaFX and a mapJFX controls [16], which display the map. The GUI might be also written using Swing and JMapViewer [17] control, but we decided to take advantage of JavaFX [18], which is a successor to Swing [19], and allows to separate the application's view layer.

Two routing services, based on OpenStreetMap [20] data, have been used. OpenRouteService API [21] is used to calculate a route from point A to B, while avoiding obstacles. Moreover, Nominatim API [22] is used to check the name of a new street, based on car position. Although a similar service is also available in the OpenRouteService API, Nominatim reverse geocoding returns better results while searching for specific streets. Note that there exist other routing service options to choose from. Instead of OpenStreetMaps we could use Google Maps, but that would require payment, in case of a large number of queries. The most popular routing services using OSM and providing a service for avoiding areas are OpenRouteService and GraphHopper [23]. We decided to take advantage of the first one, because GraphHopper does not guarantee regular routing data updates.

5 Architecture

Let us now describe the core architecture of the proposed system. The information system runs on the JADE platform and consists of three agent types. Each car is represented as an instance of a (`CarAgent`) agent. Data about blockades is sent (by `CarAgent` agent) to the central agent (`JanosikAgent` agent), which stores and processes it. Information about a new accident / traffic jams is sent (by the `JanosikAgent`) to the `DroneAgent` agent, which is responsible for checking, whether the obstacle persists or has been removed. After user defines start point and destination of travel, `CarAgent` asks `JanosikAgent` about current blockades and calculates route that omits (minimizes effects of) present obstacles. When `CarAgent` receives information about new obstacle on the road (message form `JanosikAgent`), it tries to adapt the route to avoid the problem. The main components of the proposed system have been depicted in Fig. 1.

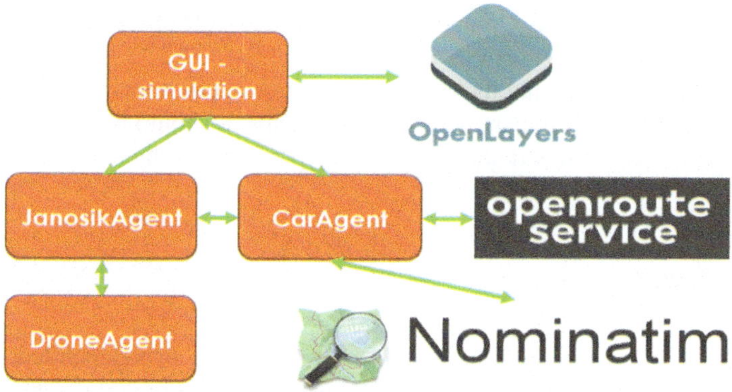

Fig. 1 System components and their interactions

Here, the three agent types and their interactions are depicted. Note that the potential bottleneck problem, in the case of the `JanosikAgent` can be easily avoided by splitting the area of interest into smaller ones (the way that Whitestein does), each overseen by a separate instance of the `JanosikAgent` agent (demarcated as responsible for a specific area). In this case, `CarAgent` agents would connect to the `JanosikAgent` agent associated with a given area, to request information about blockades. Obviously, issues related to cars moving between areas overseen by separate `JanosikAgents` would have to be dealt with, but this is only a technical problem. The `CarAgent` uses the following services: *OpenRouteService* and *Nominatim* to calculate route that avoids known obstacles, and to request its current location information. Here, the route calculation is deliberately dedicated to the `CarAgent` agent, and following the general principles of an edge-type distributed architecture (calculations should take place as close to the edge of the network as possible). Next, we see the `DroneAgent` agent, "managed" by the `JanosikAgent` agent, which assures that one of its instances (a physical drone) updates status of (a) specific road blockade(s). Finally, the GUI, which is based on OpenLayers, is depicted. Let us now describe each of system modules in more detail.

5.1 JanosikAgent

This is the central module, realized through one or more `JanosikAgent` agents. Its role is to manage information about road obstacles. It receives information from `CarAgent` agents. It also responds to them with current information needed for trip planning. Upon receiving message about new obstacle on a given street, it broadcasts it on the `JADE's Topic` corresponding to that street. This is how `CarAgents` receive obstacle information updates. Next, it requests that the `DroneAgent` controls status of given obstacle. Upon the disappearance of the obstacle, it informs cars, by sending message to the given `Topic`.

Note that the only data that the `JanosikAgent` agent stores are coordinates of obstacles, which it sends to cars and to drones. Thus, it should be able to "service" large number of vehicles, and also preserve the privacy of vehicle movement.

5.2 *CarAgent*

The `CarAgent` agent represents human user. It can be an independent software entity placed on "any device" (in case of older cars), or a module of the software running the autonomous vehicle. It completes multiple tasks in support of the traveler.

- Obtains information (from the `JanosikAgent` agent) needed to plan the route.
- Calculates the optimal route, registers to pertinent street `Topics`, and knows the current position of the car.
- Listens to messages concerning streets it is about to drive (street `Topics`).
- During travel, `CarAgent` agent de-registers from the `Topics` of streets that it has already travelled.
- When a message with a given `Topic` arrives, and indicates a new obstacle, the `CarAgent` agent asks `JanosikAgent` for updated information about obstacles in a given area (city) and calculates a new route to the destination (avoiding/minimizing effect of) obstacles.
- Informs the `JanosikAgent` agent about obstacles spotted on the road.

The `CarAgent` agent works as an "edge client." Therefore, it receives only obstacle coordinates. Then, it connects with the routing service and sends a request for a route avoiding areas around the blocked points.

As far as communication with the human is concerned, when `CarAgent` agent receives information about a new obstacle, it can display it to the driver and ask whether to change the route. However, in the implemented simulation module, the route is adjusted automatically to illustrate an instant reaction to the appearing problem.

5.3 *DroneAgent*

At this stage of system development, the `DroneAgent` agent is a virtual entity that was introduced to illustrate proposed future functionality. It receives requests from the `JanosikAgent` to check the status of an obstacle at given coordinates. When a drone arrives at the destination it undertakes one of the following actions:

- If the obstacle is no longer present, it informs `JanosikAgent`, and flies to check the status of the next obstacle on the list (or back to the base—depending on the level of fuel/energy).

- If the issue still exists, it flies to the next obstacle. The current obstacle is marked as still active and is placed at the end of the list of coordinates to be checked by this (or the next) drone.

The `DroneAgent` agent manages subsequent checks of active obstacles. Currently, only a very simplistic version of this functionality has been implemented. Obviously, as the system is further developed, `DroneAgent`'s logic and responsibilities will need to be adjusted to make it more realistic.

6 Simulation and Testing

Because testing of the proposed solution would require using cars and drones, in real traffic circumstances, we had to create a simulation module, to test system behavior in different scenarios, reflecting real situations.

The simulation module displays a JavaFX GUI. For the location, we have used map of Mexico City. Clicking on a map sets a marker on the chosen coordinates. Then, the user can (a) "create a car" in that location, (b) set a route for the car, or (c) create a obstacle at a given location. Moreover, the GUI displays routes of remaining cars, so it is possible to observe how the system works, and how the routes are adapted on the basis of materializing obstacles.

In addition, within the GUI, the user also sees the current position of the drone (in our simulation, we have used only a single drone). It is also possible to adjust the simulation speed. When it is set to *1*, the cars speed is equal to the real speed on the road received from the routing service.

Let us now describe the behavior of the system for two selected situations. While, due to space limitation, we can only present two scenarios, it has to be stressed that the implemented system has been thoroughly tested, and we are certain that it works as desired (for all, however, limited functions).

6.1 Rerouting After Obstacle Detection

Let us, first consider scenario where *Car 1* and *Car 0* travel to the northwest of the city, as presented in Fig. 2. Initially, there are no obstacles, so routes for both cars are the same.

After some time, an accident takes place in a location between *Car 1* and *Car 0*. As we can see in Fig. 3, *Car 1* does not have the obstacle in its route; hence, nothing changes. However, the accident has blocked the original route of *Car 0*, so it has received ("instant") information from the `JanosikAgent` and changed its route to avoid the blocked street.

Fig. 2 Routes of two cars without obstacles

Fig. 3 Routes after an obstacle is detected

6.2 New Obstacle Notification and Use of `DroneAgent`

Let us now illustrate what happens when `JanosikAgent` is notified about a new obstacle. Figure 4 portrays situation after car detected obstacle and notified `JanosikAgent` about it.

Here, `JanosikAgent` added obstacle to its knowledge database and asked `DroneAgent` to check obstacle's status. `DroneAgent` added obstacle to the end of its queue of coordinates that need to be visited. Since there are no other obstacles currently in the system, the drone travels to the location of the detected obstacle.

When the drone arrives to the obstacle, as depicted in Fig. 5, it checks if the obstacle is still present:

- If not, then it notifies `JanosikAgent` that the road is passable now and flies to check the next point in its queue.
- If the obstacle still exists, then it enqueues point to the end, but it will not visit this point until a certain amount of time passed after last status check for this obstacle (10 second delay was used for testing purposes).

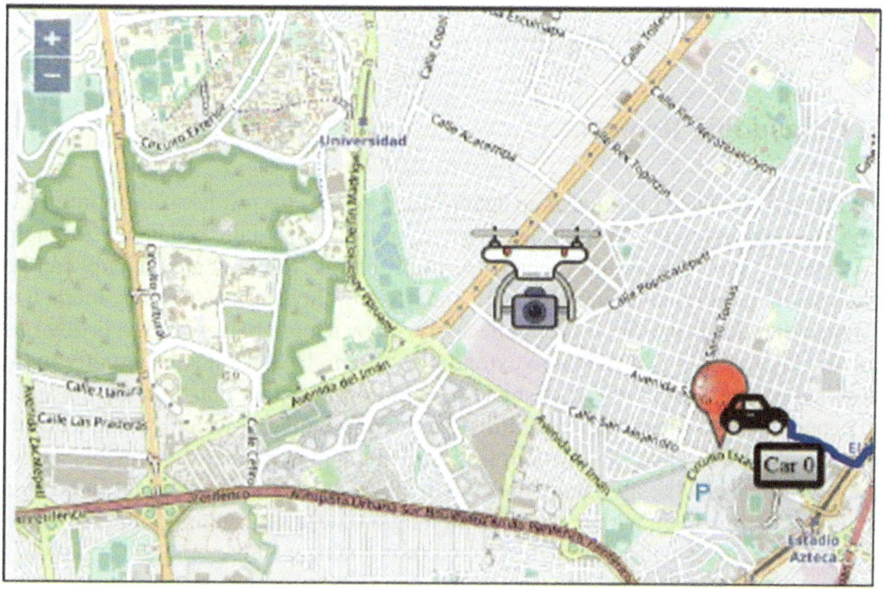

Fig. 4 Notifying system about detected obstacle

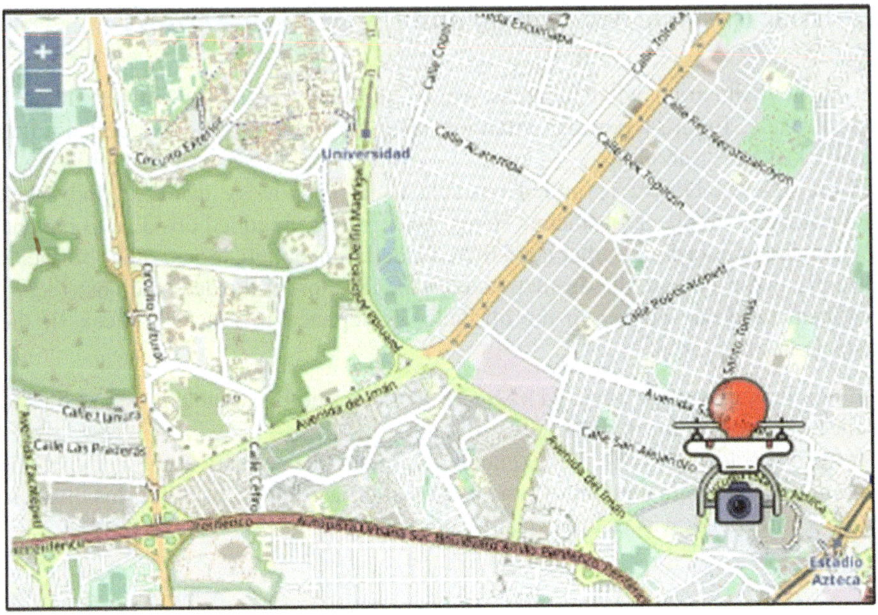

Fig. 5 Drone checking the obstacle status

7 Concluding Remarks

In this paper, we have introduced an agent-based system for traffic obstacle information detection and management. The system, in its current state, can be used to help save drivers' time during unexpected travel problems. Implemented agents can be tested using the simulation module, with maps of any chosen city. At the moment, it is assumed that the drone video should be analyzed by a human. However, it is obvious that already today, it is possible to extend the system, to make it substantially more autonomous (by adding visual obstacle recognition based on machine learning). In the paper, we have illustrated capabilities of the implemented system, using a developed simulation module. The system was implemented to be easily extensible. Here, extensions can involve all agents presented above. In particular, of great interest is the use of a drone fleet, with realistic flight length (energy consumption) parameters. We plan to proceed with system development in the near future.

References

1. Nam T, Pardo TA (2011) Conceptualizing smart city with dimensions of technology, people, and institutions. In: Proceedings of the 12th annual international digital government research conference: digital government innovation in challenging times, ACM, pp 282–291

2. About Google Maps (2019). [online], https://www.google.com/intl/pl/maps/about/. Accessed 31 Aug 2019
3. Google Maps Documentation (2019). [online], https://developers.google.com/maps/documentation/. Accessed 31 Aug 2019
4. About Waze (2019). [online], https://support.google.com/waze/?hl=en#topic=9022747. Accessed 31 Aug 2019
5. Apple Maps (2019). [online], https://en.wikipedia.org/wiki/Apple_Maps
6. Yanosik (2017). [online], https://yanosik.pl. Accessed 24 Sept 2019
7. Inrix Traffic AI (2019). [online], http://inrix.com/products/ai-traffic/
8. Constine J (2016) Inrix now collects traffic data from 100m drivers, shows black friday congestion up 32.5% despite ecommerce. [online], https://techcrunch.com/2012/11/26/inrix/
9. Kasprzyk Z, Rychlicki M, Paciorek R (2017) Proposal for implementing information systems on the road traffic conditions in Warsaw in Ochota district. technika, eksploatacja, systemy transportowe, Autobusy
10. Himoff J, Skobelev P, Wooldridge M (2005) Magenta technology: multi-agent systems for industrial logistics. pp 60–66, https://doi.org/10.1145/1082473.1082805
11. Dorer K, Calisti M (2005) An adaptive solution to dynamic transport optimization. In: Proceedings of the fourth international joint conference on Autonomous agents and multiagent systems, ACM, pp 45–51
12. Mumtaz S, Huq KMS, Ashraf MI, Rodriguez J, Monteiro V, Politis C (2015) Cognitive vehicular communication for 5g. IEEE Commun Mag 53(7):109–117
13. Jade Site (2019). [online], https://jade.tilab.com. Accessed 3 July 2019
14. Palanca J, Foundation PS (2019) spade—pypi. [online], https://pypi.org/project/spade/. Accessed 31 Aug 2019
15. Bordini RH, Hübner JF, Wooldridge M (2007) Programming multi-agent systems in agentspeak using Jason. Wiley
16. Meisch PJ (2019) mapjfx|sothawo. [online], https://www.sothawo.com/projects/mapjfx/. Accessed 5 July 2019
17. Jmapviewer (JOSM) (2019). [online], https://josm.openstreetmap.de/doc/index.html?org/openstreetmap/gui/jmapviewer/JMapViewer.html. Accessed 31 Aug 2019
18. Javafx (2016). [online], https://www.java.com/pl/download/faq/javafx.xml. Accessed 31 Aug 2019
19. Müller B (2013) Why, where, and how javafx makes sense. [online], https://www.oracle.com/technetwork/articles/java/casa-1919152.html. Accessed 31 Aug 2019
20. Openstreetmap (2019). [online], https://www.openstreetmap.org/. Accessed 5 July 2019
21. openrouteservice (2019). [online], https://openrouteservice.org. Accessed 5 July 2019
22. Nominatim (2019). [online], http://nominatim.org. Accessed 5 July 2019
23. Graphhopper (2019). [online], https://www.graphhopper.com. Accessed 31 Aug 2019

Hybridization of Constriction Coefficient-Based Particle Swarm Optimization and Chaotic Gravitational Search Algorithm for Solving Engineering Design Problems

Sajad Ahmad Rather and P. Shanthi Bala

Abstract The Constriction Coefficient-based Particle Swarm Optimization and Chaotic Gravitational Search Algorithm (CPSOCGSA) is a stochastic hybrid optimization technique. It is a combination of PSO which is inspired by the simulated behavior of bird flocking and GSA that is a physics-based heuristic technique inspired by Newton's law of universal gravitation. It combines the exploitation and exploration capabilities of PSO and GSA, respectively, to obtain the optimal result. Further, it uses ten chaotic maps for the proper balance between exploration and exploitation processes. The advantages of CPSOCGSA have been incorporated in various mechanical and civil engineering design frameworks which include Welded Beam Design (WBD), Compression Spring Design (CSD), Pressure Vessel Design (PVD), and Three Bar Truss Design (TBTD). The CPSOCGSA has been compared with seven stochastic algorithms, particularly Standard Constriction Coefficient-Based Particle Swarm Optimization and Gravitational Search Algorithm (CPSOGSA), Gravitational Search Algorithm (GSA), Classical Particle Swarm Optimization (PSO), Biogeography-Based Optimization (BBO), Continuous Genetic Algorithm (GA), Differential Evolution (DE), and Ant Colony Optimization (ACO). The experimental results indicate that CPSOCGSA was successful in minimizing the cost function and handling the constraints of the design frameworks efficiently as compared to other algorithms.

Keywords Particle Swarm Optimization · Gravitational Search Algorithm · Constriction Coefficient · Optimization · Engineering design · Chaotic maps

S. A. Rather (✉) · P. S. Bala
Pondicherry University, Puducherry 605014, India
e-mail: Sajad.win8@gmail.com

P. S. Bala
e-mail: shanthibala.cs@gmail.com

S. M. Thampi et al. (eds.), *Applied Soft Computing and Communication Networks*,
Lecture Notes in Networks and Systems 125,
https://doi.org/10.1007/978-981-15-3852-0_7

Abbreviations

GSA	Gravitational Search Algorithm
PSO	Particle Swarm Optimization
CGSA	Chaotic Gravitational Search Algorithm
CPSO	Constriction Coefficient-Based PSO
CPSOCGSA	Constriction Coefficient-Based PSO and Chaotic GSA
GA	Genetic Algorithm
DE	Differential Evolution
ACO	Ant Colony Optimization
BBO	Biogeography-Based Optimization
CPSOGSA	Constriction Coefficient-Based PSO and GSA
WBD	Welded Beam Design
CSD	Compression Spring Design
PVD	Pressure Vessel Design
TBTD	Three Bar Truss Design

1 Introduction

In the last decade, heuristic optimization algorithms are creating ripples in the international computational intelligence community of researchers. The research works, and the applications of heuristic algorithms in various fields have been phenomenal. Heuristic algorithms are getting popular due to their stochastic nature and simplicity. According to the No Free Lunch theorem (NFLT) [1], a single algorithm cannot solve all optimization problems. In simpler words, if an optimization algorithm solves some problems with high performance, there is a high probability that it performs badly in solving other types of optimization problems. Hence, the researchers have proposed many optimization algorithms to tackle different kinds of optimization issues. The famous optimization algorithms include Particle Swarm Optimization (PSO) [2] which is inspired by the social behavior of birds and fishes, Ant Colony Optimization (ACO) [3] based on the searching behavior of ants; Biogeography-Based Optimization (BBO) [4] inspired from the distribution and migration models of species, Differential Evolution (DE) [5] and Genetic Algorithm (GA) [6] which are motivated by the theory of evolution. They have been applied to solve various problems of computer science and other fields of study such as electronics, biology, and oil industry.

All nature-inspired optimization algorithms are heuristic in nature. The agents work in a parallel fashion without central control and create a collective effect called swarm intelligence. These algorithms have two basic characteristics namely exploration and exploitation. Exploration means the searching space limits of the algorithm, whereas exploitation is the process of finding the best result from a pool of feasible solutions. As per Eiben and Schippers [7], the exploration and exploitation are inversely proportional to each other, i.e., if an optimization algorithm has good exploration power for some problem, then it will be having less exploitation power for other problem. Previously, researchers were using random walks and gradient descent methods for improving exploration and exploitation, respectively. However, the aforementioned methods have the limitation of increasing the overall computational cost of the algorithm. Now, the researchers are utilizing chaotic maps for increasing diversification and local exploitation of search space to find the optimal candidate solutions [8, 9].

Chaos theory is the study of dynamic systems. The interesting property of these systems is that when there is a minor change in the system, the whole system gets affected. In simpler words, a minute change in initial parameters creates variations throughout the system. Moreover, randomness is not necessary for chaotic systems, rather deterministic systems also show chaotic behavior [10]. The optimization algorithms utilize Initial Parameter Sensibility (IPS) property of chaotic maps for fast exploitation, global exploration, and alleviation from local minima entrapment problem.

The meta-heuristic algorithms show good results on benchmark functions, but they perform poorly on real-world problems. The practical problem is a perfect benchmark for checking the problem-solving capabilities of an optimization algorithm.

To overcome the intensification, randomization, and entrapment in local minima problems; the optimization algorithms are merged, combined, and hybridized. The hybridization results in the increase of performance and accuracy of the algorithms. The researchers have used various optimization techniques such as GA, PSO, DE, GSA, and BBO for solving real-world problems in many fields such as electronic science, control systems, and operation research. Most of the algorithms are evolutionary in nature and used singly, while CPSOGSA is a recent meta-heuristic hybrid algorithm proposed by Sajad and Shanthi in 2019 [11]. Basically, CPSOGSA is a hybrid optimization algorithm that uses both PSO and GSA operators for overcoming local exploitation, randomization, and entrapment in local minima issues of standard PSO and GSA. Moreover, Constriction Coefficient is important for increasing the convergence speed of the hybrid framework. Ten chaotic maps are incorporated into standard GSA [12] and have been applied to four engineering design problems for testing its hybridization potential and complex benchmark handling capability. All these problems have unknown search spaces and are ideal for checking the problem-solving capability of the algorithm. Further, CPSOCGSA will be examined for constraint handling and parameter optimization for engineering design problems.

The paper is organized as follows. Chaotic maps are described in Sect. 2. Moreover, a brief introduction about CPSO and CGSA has been also provided in Sect. 2. Besides, hybrid CPSOCGSA is explained in Sect. 3, and the simulation results of CPSOCGSA are discussed in Sect. 4. Finally, Sect. 5 concludes the work.

2 Chaotic Maps

As already stated, chaotic maps have stochastic behavior and nonlinear motions. The chaotic behavior is not only shown by random systems but also deterministic systems as it is quite evident from Table 1. Ten chaotic maps have been used in this paper. The chaotic map equations are deterministic but have randomized fluctuations as shown in Fig. 1. All chaotic maps have particular range of values for initialization, i.e., (0, 1) and (1, −1). It should be noted that initial values have significant effects on the output of chaotic maps. So, proper selection of initial value is important for chaotic maps. For this study, chaotic maps have the same initial values which were used in Gandomi and Yang [13]. In this work, ten chaotic maps were embedded in the gravitational constant (G) of GSA. Basically, the gravitational constant is important for the proper balance between intensification and diversification phases in GSA. By incorporating chaotic maps in GSA, its exploitation and convergence speed will be affected positively.

Table 1 Chaotic maps embedded with GSA

S. No.	Chaotic function	Chaotic map	Limits
1.	Chebyshev	$l_{i+1} = \cos(i\ \cos^{-1}(l_i))$	(1, 1)
2.	Circle	$l_{i+1} = \mathrm{mod}(l_i + q - (\frac{p}{2\pi})\sin(2\pi l_k), 1)$, $p =$ 0.5 and $q = 0.2$	(0, 1)
3.	Gauss	$l_{i+1} = \begin{cases} -l, & l_i = 0 \\ \frac{1}{\mathrm{mod}(l_i, 1)}, & \text{otherwise} \end{cases}$	(0, 1)
4.	Iterative	$l_{i+1} = \sin\left(\frac{p\pi}{l_i}\right)$, $p = 0.7$	(1, 1)
5.	Logistic	$l_{i+1} = p\,l_i(1 - l_i)$, $p = 4$	(0, 1)
6.	Piecewise	$l_{i+1} = \begin{cases} \frac{l_i}{r}, & 0 \le l_i \le 0 \\ \frac{l_i - r}{0.5 - r}, & r \le l_i \le 0.5 \\ \frac{1 - r - l_i}{0.5 - r}, & 0.5 \le l_i \le 1 - r \\ \frac{1 - l_i}{r}, & 1 - r \le l_i \le 1 \end{cases}$, $r = 0.4$	(0, 1)
7.	Sine	$l_{i+1} = \frac{p}{4}\sin(\pi l_i)$, $p = 4$	(0, 1)
8.	Singer	$l_{i+1} =$ $z(7.86 l_i - 23.31 l_i^2 + 28.75 l_i^3 - 13.302875 l_i^4)$, $z = 1.07$	(0, 1)
9.	Sinusoidal	$l_{i+1} = p\,l_i^2 \sin(\pi l_i)$, $p = 2.3$	(0, 1)
10.	Tent	$l_{i+1} = \begin{cases} \frac{l_i}{0.7}, & l_i \le 0.7 \\ \frac{10}{3}(1 - l_i), & l_i \ge 0.7 \end{cases}$	(0, 1)

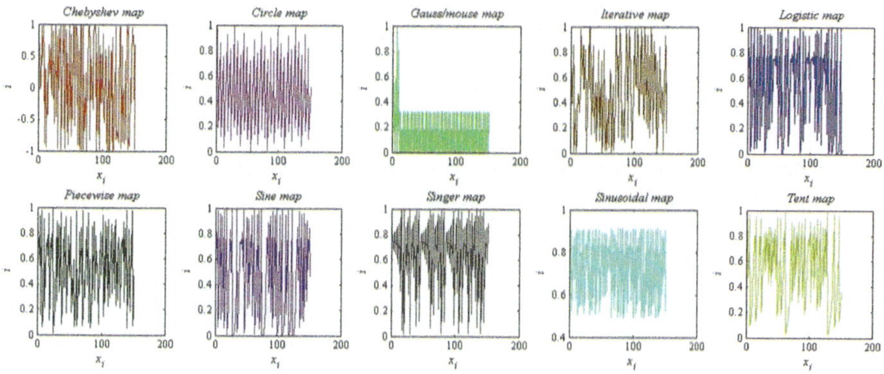

Fig. 1 Stochastic behavior of chaotic maps [8]

2.1 Constriction Coefficient-Based PSO (CPSO)

The PSO consists of three important operators which are inertia weight (w), pbest (personal particle best), and gbest (global particle best). The inertia weight gives diversification (exploration) capability and <pbest, gbest> provides intensification (exploitation) power to PSO, respectively [14]. The mathematical formulations of velocity and position of the swarm particles are given in Eqs. (1) and (2).

$$v_i^d(t+1) = w(t)v_i^d(t) + c_1 r_{i1}(\text{pbest}_i - x_i^d(\text{t})) + c_2 r_{i2}\left(\text{gbest} - x_i^d(\text{t})\right) \quad (1)$$

$$x_i^d(t+1) = x_i^d(t) + v_i^d(t+1) \quad (2)$$

whereas $<c_1, c_2>$ are cognitive and social constants and $<r_{i1}, r_{i2}>$ are the probabilistic numbers in closed interval [0, 1].

The Constriction Coefficient-Based PSO was mathematically formulated by Clerc and Kennedy [15]. Actually, they performed rigorous algebraic and eigenvalue (search procedure) analysis of the classical PSO. The main goal of CPSO is to control the particles from moving outside the search space and thus, increasing the convergence of the particles toward pbest and gbest. The various CPSO parameters and their values are shown as,

$$\begin{aligned} &\varphi_1 = 2.05, \varphi_2 = 2.05, \varphi = \varphi_1 + \varphi_2 \\ &K = 2/(\varphi - 2 + \text{sqrt}(\varphi^2 - 4) \end{aligned} \quad (3)$$

where $<\varphi, \varphi_1, \varphi_2>$ are control parameters and their values are important for controlling the particle trajectory. Further, 'K' is the Constriction Coefficient or factor which is pivotal for particle convergence toward the global optimum.

Also, when $w(t) = K$, $c_1 = K\varphi_1$ (personal learning coefficient) and $c_2 = K\varphi_2$ (global learning coefficient), the velocity Eq. (1) in modified form is shown as,

$$v_i^d(t+1) = (2/(\varphi - 2 + \text{sqrt}(\varphi^2 - 4\varphi))v_i^d(t) \\ + K\varphi_1 r_{i1}\left(\text{pbest}_i(\text{t}) - x_i^d(t)\right) + K\varphi_2 r_{i2}(\text{gbest} - x_{id}(t)) \quad (4)$$

Equation (4) shows the velocity of the particles is inversely proportional to the control parameter, φ. Also, the value of φ must be greater than 4 ($\varphi > 4$) for maintaining the stability of particle system.

2.2 *Chaotic Gravitational Search Algorithm (CGSA)*

According to modern physics, nature is composed of four forces such as the gravitational force, the strong nuclear force, the electromagnetic force, and the weak nuclear force. As per classical Newtonian mechanics, gravitational law is stated as 'the gravitational force between two masses is directly proportional to the product of their masses and inversely proportional to the square of the distance between them' [16–20].

Let $X = \{x_1, x_2, x_3 \ldots, x_n\}$ be a system with 'n' agents, such that $x_i \in R^n$. The force between mass i and j is shown in Eq. (5).

$$F_{ij} = G(t)\frac{m_{pi}(\text{t})m_{aj}(\text{t})}{R_{ij}(t) + \in}\left(x_j^d(t) + x_i^d(t)\right) \quad (5)$$

Here, $m_{pi}(t)$ is passive gravitational mass, and $m_{aj}(t)$ is active gravitational mass. $R_{ij}(t)$ is the Euclidian distance, and $\in$ is a positive constant in dth-dimensional space.

The gravitational constant 'G' is important for adjusting the accuracy of the search and is given by Eq. (6).

$$G(t) = G(t_0)\text{e}^{\left(-\alpha\frac{\text{CI}}{\text{MI}}\right)} \quad (6)$$

where $G(t)$ is the gravitational constant with respect to time interval t, α is a coefficient that decreases with time, CI is the current iteration, MI is the maximum number of iterations, and $G\ (t_0)$ indicates the initial value of the gravitational constant with a value of 20.

The gravitational constant, G has a special role to play in the proper navigation of search space. It is because G's value exponentially decreases slowly in the initial iterations that help in the exploration of the search space. Further, in the last iterations, G's value decreases with a fast exponential pace and hence, helps in the convergence of optimal candidate solutions toward heavy agents. In 2017, Mijalili et al. have introduced chaotic normalization in GSA by incorporating chaotic sequences in gravitational constant for increasing convergence rate and removing stagnation in the local minima problem of GSA. The normalization function, $C_{\text{i}}^{\text{norm}}$ with respect to interval t, is given in Eq. (7).

$$C_{\mathrm{i}}^{\mathrm{norm}}(t) = \frac{(C_i(t) - a) * (d - c)}{(b - a)} + c \tag{7}$$

where (a, b) are chaotic range values, i is the index of the chaotic map, and (c, d) are normalized intervals, while c has a value of zero. Moreover, d is given in Eq. (8).

$$d = \mathrm{MI} - \frac{\mathrm{CI}}{\mathrm{MI}}(\text{Max-Min}) \tag{8}$$

In Eq. (8), Max and Min are adaptive intervals having a value of 20 and e^{-10}, respectively. The final value of the gravitational constant is the addition of Eqs. (6) and (7).

$$G^{\mathrm{c}}(t) = C_{\mathrm{i}}^{\mathrm{norm}}(t) + G(t_0)\mathrm{e}^{\left(-\alpha\frac{\mathrm{CI}}{\mathrm{MI}}\right)} \tag{9}$$

According to Eq. (9), G^{c} has both adaptive learning capability and chaotic ergodicity. On the theoretical side, the chaotic maps have the advantages of producing abrupt changes in the value of G which allows candidate solutions to get away from local minima traps.

Moreover, the calculation of total force exerted on the system due to masses can be calculated using Eq. (10).

$$F_i^d(t) = \sum_{j=1, j\neq i}^{m} \gamma_j F_{ij} \tag{10}$$

whereas γ_j *belongs to* [0 1].

As per force law of motion, 'The acceleration (a) of the agent is directly proportional to the force (F) applied by the agent and inversely proportional to the mass (M) of the agent'. It is calculated using Eq. (11).

$$a = \frac{F}{M} \tag{11}$$

To find the global optimum, it is important to calculate the position and velocity of the heavy mass. These can be represented mathematically through Eqs. (12) and (13), respectively.

$$v_i^d(t+1) = \gamma_j v_i^d(t) + a_i^d(t) \tag{12}$$

$$x_i^d(t+1) = x_i^d(t) + v_i^d(t+1) \tag{13}$$

3 The Hybrid CPSOCGSA Algorithm

The main goal of CPSOCGSA is to use the exploration capabilities of the Gravitational Search Algorithm and convergence power of Constriction Coefficient-Based Particle Swarm Optimization. The chaotic maps are utilized for overcoming entrapment in local minima problem of standard GSA. The merging equation that combines both the algorithms is given in Eq. (14).

$$\begin{aligned} v_i^d(t+1) = (2/(\varphi - 2 + \text{sqrt}(\varphi^2 - 4\varphi))v_i^d(t) + K\varphi_1 r_{i1}\left(a_i^d(t) - x_i^d(t)\right) \\ + K\varphi_2 r_{i2}\left(\text{gbest} - x_i^d(t)\right) \end{aligned} \quad (14)$$

where v_i^d is the velocity of the swarm particles, $<\varphi, \varphi_1, \varphi_2>$ are control parameters, K is the Constriction Coefficient, a_i^d is the acceleration of the particles, and gbest represents the social capability component of the particle system.

The position of the particles is given by Eq. (15).

$$x_i^d(t+1) = x_i^d(t) + v_i^d(t+1) \quad (15)$$

The pseudo-code of CPSOCGSA is shown in Fig. 2 as follows.

The optimization algorithms are comfortable in solving continuous benchmark test functions because it is simple for agents to explore the search space and find feasible candidate solutions in these types of functions. However, the real test of intelligent techniques is solving complex nonlinear test functions like engineering benchmarks where algorithms have to deal with complex constraints and strenuous inequalities.

a) **Initialize** the search space randomly in an 'n' particle system as:

  X= $\{x_1, x_2, x_3 \ldots, x_n\}$

  V= $\{v_1, v_2, v_3 \ldots, v_n\}$.

  while (for each particle) **do**

b) **Update** the gravitational constant, G^c (t).

c) **Find** the Gravitational force, F_{ij} by using equation (5).

d) **Calculate** the acceleration, $a_i^d(t)$ with the help of equation (11).

e) **Update** the velocity: $v_i^d(t+1)$.

f) **Update** agent positions: $x_i^d(t+1)$.

  end while

 Until completing maximum number of iterations

 Return the best candidate solution

Fig. 2 Pseudo-code of CPSOCGSA algorithm

In the next section, the CPSOCGSA has been applied to four engineering design problems. It will be quite interesting to see how chaotic maps navigate the unknown search spaces of engineering benchmarks.

4 CPSOCGSA for Engineering Design Problems

Mainly, optimization algorithms are utilized to minimize the design parameter values and the overall cost of the engineering problems. The CPSOCGSA has been applied to four famous mechanical and civil engineering design problems which include Welded Beam, Compression Spring, Pressure Vessel, and Three Bar Truss frameworks. The above problems consist of both equality and inequality constraints. For handling the constraints, the penalty function method has been utilized. In the penalty method, the algorithm is penalized with high fitness function value if the optimization algorithm violates any constraint. In this paper, CPSOCGSA employs linear and scalar penalty functions.

The CPSOCGSA has been compared with CPSOGSA, Standard GSA, PSO, BBO, GA, DE, and ACO for parameter values and cost minimization of the design. For a fair comparison, all the algorithms were initialized with the same population value, number of searcher agents, and penalty constant value.

Moreover, the experiment is conducted on a system that has a 64-bit operating system, 2.20 GHz Intel @ core i5 processor, and 4 GB RAM. Further, the implementation of the algorithms and engineering design problems has been done using MATLAB R2013a. **The source codes are publicly available on GitHub, i.e.,** https://github.com/SajadAHMAD1.

The experimental results of all the algorithms were taken on a population size of 50 and maximum iterations of 500. Also, algorithms were repeated over 20 independent runs. The mean (or, average), standard deviation, and median of 20 independent trails of the best solutions are reported. Further, CPSOCGSA1 to CPSOCGSA10 represent chaotic maps in the same ascending order as shown in Table 1. Moreover, in all graphs, CPSOCGSA is represented as CCPSOGSA (i.e., chaotic version of CPSOGSA).

It is not enough to calculate the mean and standard deviation of algorithms on 20 independent trails for declaring a particular algorithm as an efficient one [21]. Rather, statistical tests should be conducted for a fair comparative analysis of evolutionary algorithms. In this paper, a nonparametric statistical test, namely Wilcoxon rank-sum test, has been carried out at a 5% significance level. When the particular algorithm has the minimum mean and standard deviation values, it will not be compared with itself and represented in tables by N/A, i.e., 'not applicable.' In simpler words, it is the best-performing algorithm. In addition, 'NN' symbol is used for chaotic maps which are not considered for statistical test because of having sub-optimal values for cost function of the design problem. This terminology will be followed throughout the paper. Also, the best cost values are highlighted in a bold font, and p values greater than 0.05 indicate the competitiveness of the particular algorithm as its cost function values are statistically significant than the best one.

Table 2 Initialization parameters of algorithms

Algorithm	Name of the parameter	Value of the parameter
GSA	Elitist check (Number of fittest agents after stopping criterion)	1
	R_{power} (Power of distance between agents)	1
	Min_flag (1: Minimum; 0: Maximum)	1
ACO	Pheromone update constant	1
	Initial pheromone	10
	Pheromone sensitivity	0.3
	Visibility sensitivity	0.1
GA	P_c (Crossover probability)	0.95
	P_M (Mutation probability)	0.001
	Er (Elitism)	0.2
DE	Lower bound of scaling factor	0.2
	Upper bound of scaling factor	0.8
	PCR (Crossover probability)	0.8
PSO	$C1$, $C2$ (Personal and social constants)	2
	W_{max} (Maximum inertia weight)	0.9
	W_{min} (Minimum inertia weight)	0.2
BBO	n_{Keep} (Number of habitats retained after every generation)	0.2
	$P_{mutation}$ (Mutation probability)	0.9
CGSA	Chaotic initial value	20

All optimization algorithms are first initialized before going through the various phases of iterations. They have different parameters that should be first initialized. Table 2 shows the initial parameter values of 7 participating algorithms in this work. The initial values of all algorithms are according to their traditional default versions, i.e., values used by the authors who have proposed the algorithms. Let us take the example of PSO. Table 2 depicts $C1$, $C2$, W_{max}, and W_{min} parameters of PSO having values of 2, 2, 0.2, and 0.2, respectively. Every researcher in the swarm optimization field knows that personal and social constants of PSO have a value of 2 because it shows more suitable results on these values. Likewise, parameters for other algorithms were selected considering the stability and fairness of the experimental setup.

4.1 Welded Beam Design (WBD) Problem

It is basically a minimization problem in which an algorithm is used to reduce the manufacturing cost of the design [22, 23], as illustrated in Fig. 3.

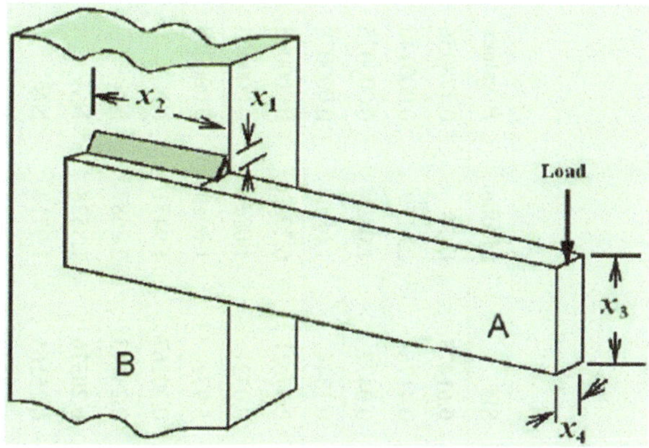

Fig. 3 Welded Beam Design problem [22]

There are four variables involved which are the length (l), height (t), thickness (b), and weld thickness (h) of the bar. It consists of several constraints such as shear (τ) and beam blending (θ) stress, bar buckling load (P_c), beam end deflection (δ), and side constraints.

4.1.1 Simulation Results and Discussion

Table 3 provides the comparative analysis of ten chaotic map versions of CPSOGSA with classical GSA, PSO, BBO, GA, DE, and ACO for the WBD problem. It is clear from Table 3 that CPSOCGSA3 (Gauss map) has shown the best results for the cost function of the WBD problem as compared to other participating algorithms and chaotic maps. It also depicts minimum values for mean and standard deviation (SD). Moreover, the p values indicate that CPSOGSA is statistically significant as compared to CPSOCGSA3 as it has p values greater than 0.05. Also, the PSO, BBO, GA, and DE provide large values for mean and SD depicting sub-optimal performance.

The comparative analysis of CPSOCGSA (equivalent to CPSOCGSA3 as it has minimum average value) with seven other algorithms is depicted in Fig. 4. It can be seen that CPSOCGSA3 is having irregular values at the start of the iteration phase, but afterward the fitness values of the solutions remain at minimum best values. Further, PSO and DE also show good exploitation capability as their convergence curves are exponentially decreasing showing lower cost function values at consecutive runs.

4.2 Compression Spring Design (CSD) Problem

This is another classical mechanical engineering problem. The cost function comprises of calculating the weight of the tension spring [24], as illustrated in Fig. 5.

Table 3 Experimental results of WBD problem

Algorithm	h	l	t	b	Best	Worst	Mean	SD	Median	P values
CPSOGSA	0.1596	4.3177	9.0366	0.2057	1.7194	2.3254	**1.927**	**0.61475**	1.888	0.178956
GSA	0.1795	6.8583	10	0.2441	1.8868	3.7099	2.8718	0.41761	2.8179	0.000140
PSO	0.7842	2	2	2	1.08el4	1.08el4	1.08el4	0.0477	1.08el4	0.000075
BBO	0.94498	1.6292	2	2	1.08el4	1.08el4	1.08el4	0.1344	1.08el4	0.000078
GA	0.2077	0.1153	0.4919	0.1711	8.52e18	1.16e22	1.96e21	2.97e21	6.20e20	0.000088
DE	0.7852	2	2	2	1.08el4	1.08el4	1.08el4	0.0321	1.08el4	0.000063
ACO	1	4	3	2	1.69e05	1.69e05	1.69e05	5.97e−11	1.69e05	0.000088
CPSOCGSA1	0.2134	3.8146	8.3799	0.2431	1.7645	2.4010	2.0204	0.20267	1.9451	NN
CPSOCGSA2	0.1627	4.3748	8.8280	0.2318	1.8999	2.6483	2.2236	0.23143	2.2687	NN
CPSOCGSA3	0.1307	9.8507	9.4640	0.2064	1.7395	2.4642	**1.9114**	**0.20576**	1.8528	N/A
CPSOCGSA4	0.1335	10	10	0.2047	1.787	2.5601	2.0567	0.25165	1.9714	NN
CPSOCGSA5	0.1665	7.9830	8.9796	0.2084	1.7469	2.4986	2.0563	0.22242	2.0032	NN
CPSOCGSA6	0.1914	10	9.0936	0.2058	1.9019	2.6119	2.1871	0.22042	2.147	NN
CPSOCGSA7	0.1355	6.8163	10	0.2062	1.7568	2.5255	2.0913	0.19994	2.0426	NN
CPSOCGSAS	0.2147	6.4022	8.0688	0.2593	1.9034	2.6648	2.2635	0.24406	2.2832	NN
CPSOCGSA9	0.2095	3.4752	9.1395	0.2260	1.8914	2.7084	2.2643	0.28617	2.2069	NN
CPSOCGSA10	0.2585	2.9933	7.9852	0.2671	1.7841	2.4746	2.0044	0.21257	1.9334	NN

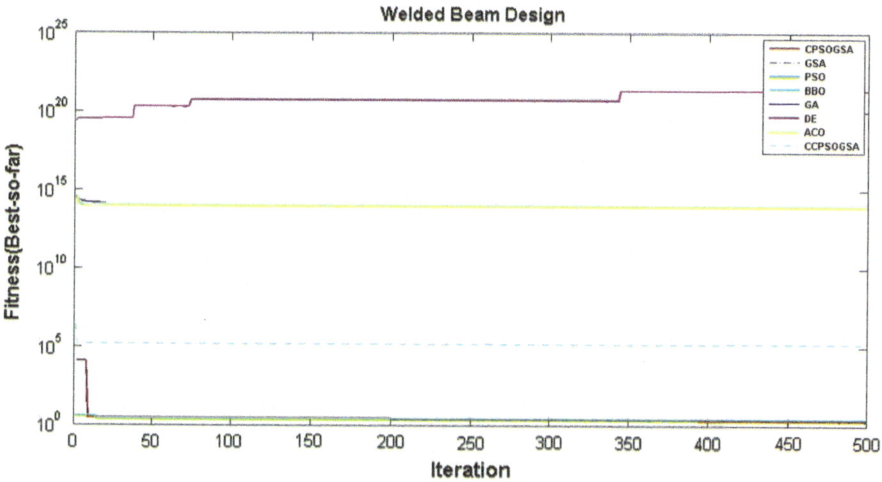

Fig. 4 Convergence curve of WBD problem

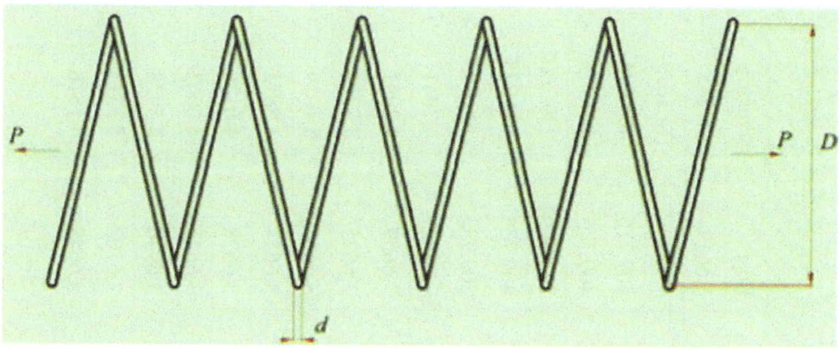

Fig. 5 Compression Spring Design problem [24]

There are three variables involved which are the diameter of the wire (d) and coil (D), respectively, and number of active coils (N).

4.2.1 Simulation Results and Discussion

It is clear from Table 4 that CPSOCGSA3 (Gauss map) provides minimum values for mean and standard deviation as far as chaotic maps are concerned. Further, CPSOGSA also shows optimal performance for the minimization of the cost function for the CSD problem as it gives minimum values for mean and standard deviation. However, the Wilcoxon rank-sum test shows that CPSOCGSA3 is statistically significant as compared to CPSOGSA as it has p values greater than 0.05.

Table 4 Experimental results of CSD problem

Algorithm	*d*	*D*	*N*	Best	Worst	Mean	SD	Median	*P* values
CPSOGSA	0.1391	1.3000	11.8924	3.6619	3.7245	**3.671**	**0.017737**	3.6619	N/A
GSA	0.1074	1.3000	6.9467	4.425	14.1219	8.1426	2.9629	6.8761	0.000088
PSO	2	2	2	409.775	409.775	409.775	2.33e−13	409.775	0.000088
BBO	2.002	2	2	409.775	409.775	409.775	0.0600	409.775	0.000088
GA	0.6917	0.6919	14.6698	283.9094	2396.001	479.338	474.9263	314.3773	0.000088
DE	2	2	2	409.775	409.775	409.775	2.332 − 13	409.775	0.000088
ACO	2	1	3	209.933	209.933	209.933	5.83E−14	209.933	0.000088
CPSOCGSA1	0.1415	1.3000	12.4460	3.6652	4.7041	3.8727	0.279	3.7723	NN
CPSOCGSA2	0.1400	1.2661	12.7920	3.6659	4.2097	3.8184	0.17045	3.7547	NN
CPSOCGSA3	0.1318	1.1214	14.9020	3.6625	3.7724	**3.6948**	**0.034963**	3.6781	0.178900
CPSOCGSA4	0.1399	1.3000	12.1356	3.6665	5.1019	3.8675	0.33219	3.761	NN
CPSOCGSA5	0.1382	1.2273	13.4931	3.6654	3.9871	3.7595	0.091892	3.7512	NN
CPSOCGSA6	0.1398	1.3000	12.3433	3.6628	4.6136	3.8679	0.21931	3.8045	NN
CPSOCGSA7	0.1338	1.1553	15	3.6647	4.0063	3.8207	0.12098	3.7752	NN
CPSOCGSA8	0.1267	1.0798	14.2777	3.6936	4.5928	3.8939	0.22936	3.8059	NN
CPSOCGSA9	0.1297	1.1693	13.5533	3.6702	5.3313	4.1171	0.47001	3.9479	NN
CPSOCGSA10	0.1350	1.1637	14.3862	3.6648	4.1878	3.8036	0.14193	3.7577S	NN

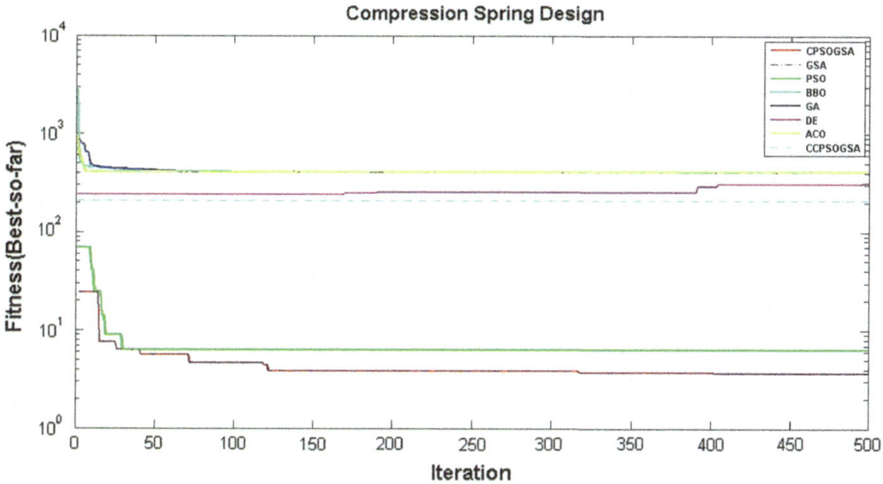

Fig. 6 Convergence curve of CSD problem

In addition, the PSO, BBO, GA, DE, and ACO have large values for the cost function of the CSD problem. It implies that they have less exploration and exploitation power in finding optimal candidate solutions in the complex search space.

The comparative analysis of CPSOCGSA (equivalent to CPSOCGSA3 as it has minimum average value) with seven other algorithms is depicted in Fig. 6. It can be seen that CPSOCGSA, CPSOGSA, and DE show good convergence power as their curves are at the bottom of Fig. 6 depicting high intensification capability in finding the global minimum.

4.3 *Pressure Vessel Design (PVD) Problem*

The configuration of the PVD framework [25] is shown in Fig. 7. The pressure vessel has capped ends and hemispherical heads. There are three variables involved which are shell thickness (T_s), the thickness of head (T_h), radius (R), and section length of the vessel (L).

Fig. 7 Pressure Vessel Design problem [25]

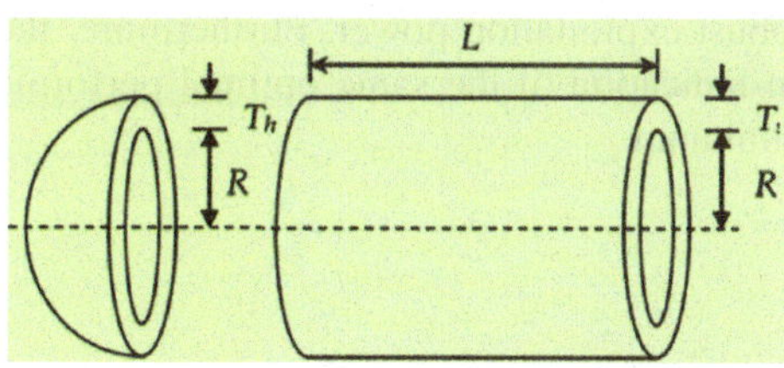

4.3.1 Simulation Results and Discussion

The thorough analysis of Table 5 shows that CPSOCGSA8 (Singer map) has minimum values for mean and standard deviation. Moreover, from the other seven optimization algorithms, only CPSOGSA has competitive values for the cost function of the PVD problem. The *p* values also agree with the significant superiority of CPSOCGSA8 as they are greater than 0.05. The sub-optimal performance of PSO, BBO, GA, and DE can be easily recognized from their large values for mean and SD measures.

Furthermore, Fig. 8 shows the convergence curves of all the algorithms at 500 iterations. It is obvious that CPSOCGSA8 has a flat curve at the bottom which conveys that it has similar best cost function values at consecutive iterations. Moreover, BBO and GA have exponentially increasing curves depicting large values for the cost function.

4.4 *Three Bar Truss Design (TBTD) Problem*

The schematic diagram of the TBTD [26] problem is shown in Fig. 9. It is a special case of fractional programming as highly nonlinear constraints are involved in the problem. The cost function of the TBTD framework is to minimize the weight of the truss.

4.4.1 Simulation Results and Discussion

Table 6 shows that PSO provides optimal performance for the minimization of the truss weight. Further, if we analyze the mean SD and best values, it can be seen that DE has similar statistical results as that of PSO. This interpretation of results is amazingly shown by Wilcoxon rank-sum test. Because the *p*-value of DE is 1 (that is, greater than 0.05), which means DE has same performance as that of PSO or the statistical results of PSO are not statistically significant as compared to DE. Further, CPSOCGSA7 (sine) also provides appreciable mean and SD values for the cost function of the TBTD problem.

Moreover, Fig. 10 depicts that DE and PSO have the same shape of convergence curves. Also, the best values of all other algorithms except GA and DE have robust exploitation power. Furthermore, the overlapping of convergence curves is an indication of the same optimal performance of an algorithms in finding global minimum.

Table 5 Experimental results of PVD problem

Algorithm	T_s	T_h	R	L	Best	Worst	Mean	SD	Median	P values
CPSOGSA	0.6392	0	46.9452	124.59	3263.11	3638.73	**3586.73**	**86.3045**	3624.92	0.016880
GSA	0.7448	0	51.7061	96.7094	3362.38	3762.82	3542.62	92.90	3531.89	0.033340
PSO	10	10	54.1477	68.5034	2.04305	2.04e05	2.04e05	7.4842	2.04e05	0.000088
BBO	10	10	54.9803	63.1639	2.04e05	2.12e05	2.06e05	2.35e03	2.05e05	0.000088
GA	66.846	73.723	17.1417	24.6301	1.56el6	1.65el6	1.62el6	2.54el4	1.63el6	0.000088
DE	IO	10	53.6320	71.9220	2.04e05	2.04e05	2.04e05	13.4617	2.04e05	0.000088
ACO	1	4	2	3	1.67el6	1.67el6	1.67el6	0	1.67el6	0.000073
CPSOGSA1	1.0893	0	65.2304	10	2303.21	3644.08	2991.36	651.06	3243.639	NN
CPSOGSA2	10900	0	65.2340	10	2303.52	3540.80	2958.75	670.87	2867.04	NN
CPSOGSA3	1.0919	0	65.2306	10	2302.68	3632.34	3018.63	664.96	3515.73	NN
CPSOGSA4	1.0953	0	65.2396	10	2303.75	3641.99	3294.69	586.94	3627.86	NN
CPSOGSA5	1.0937	0	65.2322	10	2302.73	3641.57	2806.09	637.98	2305.03	NN
CPSOGSA6	0.4740	0	40.7825	194.12	2304.20	3641.98	3058.98	638.79	3384.71	NN
CPSOGSA7	1.0824	0	65.2305	10	2303.98	3638.58	3085.82	617.11	3499.42	NN
CPSOGSA8	1.1000	0	65.2510	10	2303.24	3644.45	**2771.39**	**644.40**	2310.07	N/A
CPSOGSA9	0.4477	0	40.5625	196.72	2303.01	3649.07	3248.22	592.57	3628.25	NN
CPSOGSA10	0.4573	0	40.3555	200	2303.06	3639.39	3032.55	675.51	3607.65	NN

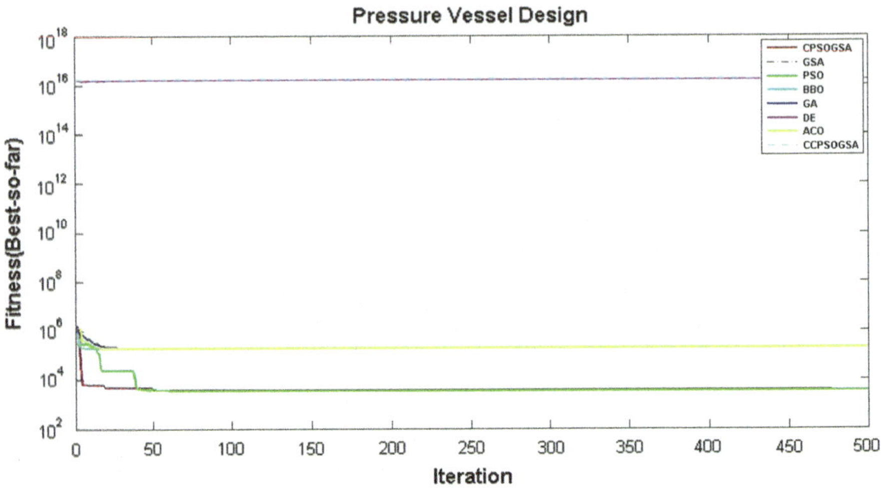

Fig. 8 Convergence curve of PVD problem

Fig. 9 Three Bar Truss problem [26]

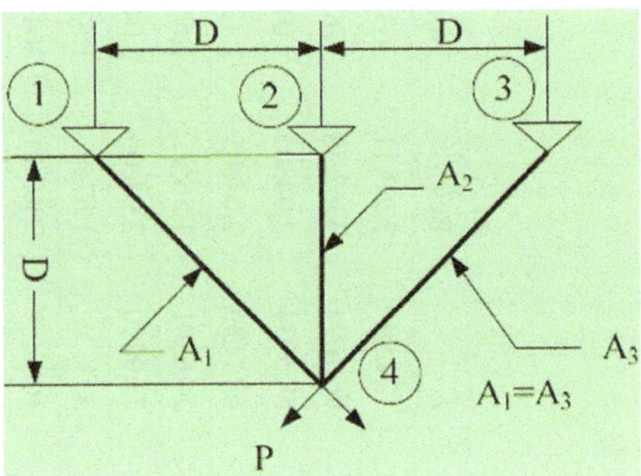

5 Conclusion

In this paper, Constriction Coefficient-Based PSO and Chaotic GSA (CPSOCGSA) is hybridized and has been applied to four engineering design problems. The chaotic map versions of the hybrid framework have shown efficient performance on three of the four design problems, especially welded beam, compression spring, and Pressure Vessel Design problems. From the ten chaotic maps, Gauss map was optimal in showing the best results to cost functions of the WBD and CSD design problems.

Generally speaking, CPSOCGSA has been quite successful in overcoming convergence issues of standard CPSOGSA. Moreover, stochastic sequences of chaotic maps were superior to random number generators because they efficiently took agents from sub-optimal regions of local minima which in turn helps in the fast convergence of the hybrid CPSOCGSA algorithm.

Table 6 Experimental results of TBTD problem

Algorithm	l_1	l_2	Best	Worst	Mean	SD	Median	P values
CPSOGSA	0.7898	0.4022	186.9465	200	192.6104	4.6741	190.8126	0.000086
GSA	0.8299	0.2794	188.1522	200	193.9115	3.7944	194.3463	0.000086
PSO	0.7868	0.2880	186.3859	186.3859	**186.3859**	**5.79e−14**	186.3859	N/A
BBO	0.7879	0.2859	186.3859	186.497	186.3955	0.026692	186.3866	0.000088
GA	0.0039	0.0981	1.521e07	1.74e13	8.819e11	3.896e12	5.105e08	0.000085
DE	0.7868	0.2880	186.3859	186.3859	186.3859	4.299e−06	186.3859	1
ACO	1	2	400	400	400	0	400	0.000007
CGSACPSO1	0.8629	0.2090	187.6367	200	193.7797	4.0957	193.5011	NN
CGSACPSO2	0.7398	0.4612	188.5264	200	194.8829	3.677	194.7421	NN
CGSACPSO3	0.7559	0.3597	187.1544	200	194.7552	4.6669	194.3901	NN
CGSACPSO4	0.8320	0.2180	187.6711	200	195.6874	4.2873	196.2945	NN
CGSACPSO5	0.7732	0.4078	186.9844	200	193.2474	3.8012	192.48	NN
CGSACPSO6	0.7836	0.3952	186.7698	200	193.9986	5.1404	195.7579	NN
CGSACPSO7	0.8236	0.2248	186.5037	200	**192.1546**	**4.6635**	191.4086	0.000088
CGSACPSO8	1	0	187.6118	200	192.9315	4.4594	191.13	NN
CGSACPSO9	0.9484	0.0636	187.0098	200	192.5639	4.0629	191.6389	NN
CGSACPSO10	0.7172	0.4726	186.8486	200	193.4458	4.0965	193.3401	NN

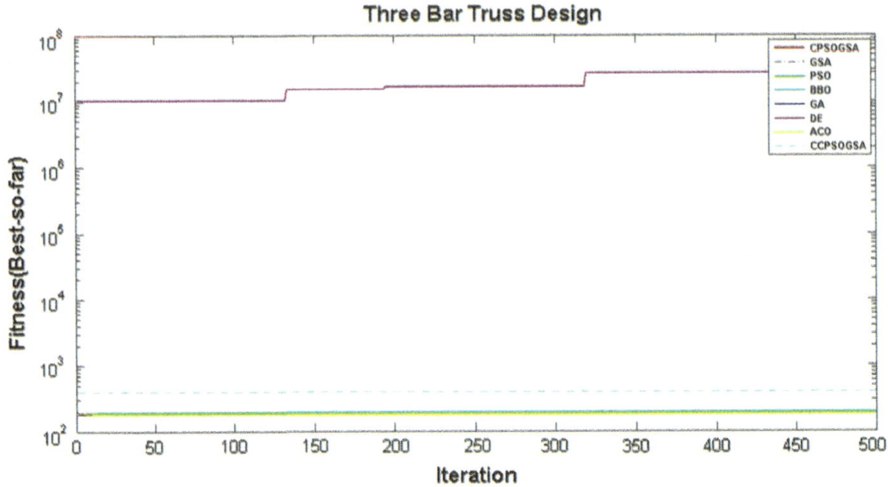

Fig. 10 Convergence curve of TBTD problem

References

1. Wolpert DH, Macready WG (1997) No free lunch theorems for optimization. IEEE Trans Evol Comput 1:67–82
2. Kennedy F, Eberhart RC (1995) Particle swarm optimization. Proc IEEE Int Conf Neural Netw 4:1942–1948
3. Dorigo M, Maniezzo V, Colorni A (1996) The ant system: optimization by a colony of cooperating agents. IEEE Trans Syst Man Cybernet -Part B 26(1):29–41
4. Simon D (2008) Biogeography based optimization. IEEE Trans Evol Comput 12(6):702–713
5. Storn R, Price KV (1995) Differential evolution—A simple and efficient adaptive scheme for global optimization over continuous spaces. Technical report TR-95-012. International Computer Science Institute
6. Tang KS, Man KF, Kwong S, He Q (1996) Genetic algorithms and their applications. IEEE Sig Process Mag 13(6):22–37
7. Eiben AE, Schippers CA (1998) On evolutionary exploration and exploitation. Fundam Inf 35:35–50
8. Mirjalili S, Gandomi AH (2017) Chaotic gravitational constants for the gravitational search algorithm. Appl Soft Comput 53:407–419
9. Gandomi AH, Yun GJ, Yang XS, Talatahari S (2012) Chaos-enhanced accelerated particle swarm optimization. Commun Nonlinear Sci Numer Simul 18:89–98
10. Kellert SH (2017) In the wake of chaos: unpredictable order in dynamical systems. University of Chicago Press
11. Rather SA, Bala PS (2019) Hybridization of constriction coefficient based particle swarm optimization and gravitational search algorithm for function optimization. In: International conference on advances in electronics, electrical and computational intelligence. IIIT Allahabad (Accepted for publication)
12. Rashedi E, Nezamabadi-pour H, Saryazdi S (2009) GSA: a gravitational search algorithm. Inf Sci 179:2232–2248
13. Gandomi AH, Yang XS (2014) Chaotic bat algorithm. J Comput Sci 5:224–232
14. Shi Y, Eberhart RC (1998) A modified particle swarm optimizer. In: IEEE international conference on evolutionary computation, pp 69–73

15. Clerc M, Kennedy J (2002) The particle swarm—Explosion, stability, and convergence in a multidimensional complex space. IEEE Trans Evol Comput 6:58–73
16. Halliday D, Resnick R, Walker J (2000) Fundamentals of Physics, 6th edn. Wiley, USA
17. Rather SA, Bala PS (2019) Analysis of gravitation based optimization algorithms for clustering and classification. Handbook of research on big data clustering and machine learning. IGI Global, pp 77–99. https://doi.org/10.4018/978-1-7998-0106-1.ch005
18. Rather SA, Bala PS (2019) A holistic review on gravitational search algorithm and its hybridization with other algorithms. In: 2019 IEEE International Conference on Electrical, Computer and Communication Technologies (ICECCT), pp 1–6. https://doi.org/10.0.1109ICECCT0.20190.8869279
19. Rather SA, Sharma N (2017) GSA-BBO hybridization algorithm. Int J Adv Res Sci Eng 6(2):596–608
20. Rather SA, Shahid M, Bala PS (2019) A comprehensive survey on solving clustering and classification problems using gravitational search algorithm. In: 2019 IEEE 9th International Conference on Advanced Computing (IACC). Tiruchirappalli, India, pp 13–18. https://doi.org/10.1109/IACC48062.2019.8971589
21. Derrac J, García S, Molina D, Herrera F (2011) A practical tutorial on the use of nonparametric statistical tests as a methodology for comparing evolutionary and swarm intelligence algorithms. Swarm Evol Comput 1:3–18
22. Coello CA (2000) Application of penalty function method in solving design problems. Comput Ind Eng 113–127
23. Rather SA, Bala PS (2020) Swarm-based chaotic gravitational search algorithm for solving mechanical engineering design problems. World J Eng 17(1):97–114
24. Arora JS (2000) Introduction to optimum design. Academic Press, US
25. Sandgren E (1990) NIDP in mechanical design optimization. J Mech Des 112(2):223–229
26. Basset MA, Wang GG, Sangaiah AK, Rushdy E (2019) Hybridization of KH and CS algorithms for solving design problems. Multimedia Tools Appl 78(4):3861–3884. https://doi.org/10.1007/s11042-017-4803-x

Parallel Association Rules Pruning Algorithm on Hadoop MapReduce

Mohamed A. Alasow, Salahadin A. Mohammed, and El-Sayed M. El-Alfy

Abstract Mining association rules is essential in the discovery of knowledge hidden in datasets. There are many efficient association rule mining algorithms. The problem is with the large number of rules they often discover. Large number of rules makes the discovery of knowledge very challenging because too many rules are difficult to understand, interpret, or visualize. To reduce the number of discovered rules, researchers proposed a number of solutions. However, these solutions are limited to the rules generated from traditional datasets and are incapable of handling rules generated from big datasets. To solve this problem, this paper proposes a Hadoop MapReduce-based parallel association rule pruning algorithm, named PPrune. Experimental results show that PPrune to be efficient and has good speedup, scaleup, and sizeup.

Keywords Data mining · Knowledge discovery · Association rules · Clustering · Hadoop MapReduce · Pruning

1 Introduction

Association rule mining (ARM) is a branch of data mining used to discover frequent patterns, associations, correlations, and other relationships between data items of a dataset [9]. In other words, ARM aims to find rules which enable us to predict the occurrence of a specific data item based on the occurrences of other data items in a given dataset. An example of an association rule (AR) would be if a customer buys bread, he is 75% likely to also buy milk. An AR is of the form $X \Longrightarrow Y$, where X and Y are sets of data items and $X \cap Y = \emptyset$.

M. A. Alasow · S. A. Mohammed · E.-S. M. El-Alfy (✉)
King Fahd University of Petroleum and Minerals, Dhahran, Saudi Arabia
e-mail: alfy@kfupm.edu.sa

M. A. Alasow
e-mail: g201409240@kfupm.edu.sa

S. A. Mohammed
e-mail: adam@kfupm.edu.sa

S. M. Thampi et al. (eds.), *Applied Soft Computing and Communication Networks*, Lecture Notes in Networks and Systems 125,
https://doi.org/10.1007/978-981-15-3852-0_8

ARM has many applications in marketing, medicine, finance, security, weather, bio-informatics, airline information systems [17], and many other fields. Due to the above-mentioned applications and many others, it has attracted many researchers to propose different ARM algorithms that extract rules from datasets [2].

Often, the number of strong rules discovered by ARM algorithms is large, even from a moderately sized dataset. Large number of rules are difficult to inspect, analyze, visualize, or interpret, which reduces the usefulness of the discovered rules in decision making and other applications [1].

To address the problem of large number of rules, different solutions have been proposed in the literature [2, 5, 8, 13, 19]. Some of the proposed solutions include constraint-based mining, meta-rules, rules pruning, and rules grouping or clustering. But all the existing solution are limited to traditional datasets. No one has proposed a solution to deal with the huge number of rules generated from big datasets. Motivated by this, in this paper, we are proposing, to the best of our knowledge, the first Hadoop and MapReduce-based rules pruning algorithm that prunes rules based on their structure and that can handle huge volume of rules discovered from big datasets.

The rest of this paper is organized as follows. Section 2 briefly reviews the published work related to our work. Section 3 describes the proposed algorithm. Section 4 presents experimental results and analysis. Section 5 draws the conclusion of our work and suggests some future work.

2 Related Work

Several approaches have been proposed in the literature to solve the problem of huge association rules generated by ARM algorithms. Summarizing and pruning solutions have been suggested to remove uncertainty and redundant rules. Aijun et al. [2] introduced two domain knowledge-based methods to prune and summarize generated rules. In the first method, to group rules, the user defines the semantic distance between rules using data taxonomy. The second method groups all rules that share an item in the antecedent and consequent.

In [16], the idea of domain ontologies is introduced to generalize association rules. This concept facilitates the representation of mined rules in the form of (an is-a hierarchy). The authors proposed post-processing method to prune and filter association rules. They integrate user knowledge modeled in the form of ontology connected to the data to improve the selection of interesting rules.

Torvonin et al. [19] used a rule cover method to differentiate between the redundant rules from non-redundant. Brijs et al. [6] enhanced the rule cover method using integer programming techniques to maximize rule redundancy reduction amount in the generated rules. Although these approaches help to prune interesting rules, they depend on domain knowledge to give users the ability to view interesting rules and discard the uninteresting rules.

Unlike domain knowledge-based approaches, various subjective and objective interesting measurements are introduced to segregate all interesting rule based on a given measure from the rest of rules. Bayardo et al. [4] came up with minimum improvement concept to perform prune interesting rules. They measured the difference between the confidence of rule and that of its proper sub-rules is measured to take the rule pruning action. In this method, choosing a low-minimum improvement threshold can lose sensitivity to catch many overlapping rules.

Bay and Pazzani [3] considered contrast sets contain the conjunction of meaningfully different attributes and values across their distribution in the set of classes or groups of interest to prune and remove insignificant contrast rules. Huang and Webb [12] anatomically discard the discovered insignificant rules in the discretized quantitative attributes using search algorithm called OPUS. Liu et al. [15] utilized chi-square statistical test to prune the rules, and the authors evaluated the dependence of antecedent and consequence of a rule by chi-square test. To make pruning of the rule result set, he used a pre-specified threshold value compared with the chi-square test. This method suffered from the issue of data sparsity to prune many rules.

Chawla et al. [7] suggested an adaptive local pruning method using directed hypergraph to prune singleton consequent association rules. In this method, the process of pruning is applied locally to keep the importance of non-local rules separate from the pruning process. However, this method fails to maintain the global picture of association rules.

Visualization-based techniques such as parallel coordinate plots [20], matrix-based visualizations [10] are introduced as post-processing techniques to analyze the discovered association rules. These techniques help to visualize the interrelations between association rule categories in a great detail. Unfortunately, most visualization techniques cannot display large sets of rules.

Classification approaches have been practiced to classify association rules. Based on classification classes CBA, Liu et al. [14] created a class attribute constraint to control the consequent of rules before generating from association rules. Similarly, a novel method is introduced in [14] to reduce the generated association classes using rule consequent constraint based on (NECR-tree). An efficient classification based on multiple association rules (CMAR) replaced to classify association rules. CMAR finds rules from prefix tree structure like CR-tree CMAR using FPgrowth Han et al. [11] and that leads to finding more specific and lower-level rules than CBA.

Clustering rules methods are also employed as an alternative way to group discovered association rules by some researcher. The clustering-based methods usually draw measuring distance between association rules to carry association rule clustering. For example, Toivonen et al. [19] assumed that the distance of association rules can be estimated by calculating how many rules are different in the overall number of rows to group association rules. Gupta et al. [18] improved the distance metric method by normalization to cluster the rules. This approach is restricted to only two attribute antecedents. The distance between two rules using conditional probability is estimated. They introduce a special case of the agglomerative chain algorithm named Dimensionless Agglomerative Chain Clustering to cluster association rules. The proposed algorithm allows clustering without knowing dimension information. Moreover, they utilize self-organizing feature map (SOM) to cluster points into a Euclidean space, and then, they apply visualization technique.

Algorithm 1 The Map method of PPrune

```
1: procedure MAP(key, r)
2:     r.consequent ← get-consequent(r)
3:     r.antecedent ← get-antecedent(r)
4:     r.size ← get-size(r.antecedent)
5:     r.antecedent ← sort(r.antecedent)
6:     k1 ← r.consequent
7:     v1 ← r.size + "," + r;
8:     Emit(k1, v1);
9: end procedure
```

3 The Proposed Rule Structure-Based Pruning Algorithm

Rule structure-based pruning is a fast way of reducing the number of strong association rules. In its conservative form, it prunes the rules without reading the transaction database. The idea is based on structural rule cover which is explained in [19]. A *rule cover* states that if X, Y, and Z are three itemsets of a given transactional database D, then $\Phi(X, Y, Z) \subseteq \Phi(X, Y)$, where $\Phi(x)$ is set of transactions in D that match the itemset x. In other words, the transactions that match the rule $X, Z \Rightarrow Y$ are a subset of the transactions that match $X \Rightarrow Y$. If rules such as $X, Z \Rightarrow Y$ are removed from a rule cover, then the remaining set which contains $X \Rightarrow Y$ is a rule cover. A structural rule cover consists of the most general rules of the original set of rules.

The proposed parallel association rules pruning algorithm, which we named PPrune, is implemented using MapReduce and runs on Hadoop. For a given set of rules R, the Map method of PPrune reads each $r \in R$ and identifies its antecedent and consequent, r.antecedent and r.consequent. It then sorts r.antecedent and computes its size, *r*.size, which is the number of items in *r*.antecedent. At last, it makes *r*.consequent as the key and *r*.antecedent and *r*.size as the value and sends them to the reducer.

The Map method of PPrune is depicted in Algorithm 1. It takes r as input and computes its antecedent, its consequent, and the size of its antecedent, lines 3–5. It then makes the consequent as the key and the antecedent and its size as the value and sends them to the reducer, lines 6–8.

The Reduce method of PPrune takes as input a rule consequent as a key, $K1$, and set of strong association rules with the same consequent as a value, $V1$. It gives as output a set of general rules as output value, $V2$. It goes through two phases. In Phase 1, it puts each r in $V1$ with the same r.consequent and r.size into the same sub-partition, S[*r*.size], lines 2–4. After it puts each rule in one of the sub-partition of S, it deletes the empty sub-partitions of S by resizing and re-indexing S, Line 5.

In Phase 2, PPrune prunes covered rules and starts at Line 6. First it initializes the set of general rules G to null, Line 6. In lines 7–15, it loops through each sub-partitions of S and each rule in a sub-partition and does the following. If a rule belongs to the first sub-partition, $S[1]$ or the rule is not a superset of any of the rules in G, then the rule is added to G, otherwise it is pruned. At last, at Line 18, it emits G.

Algorithm 2 The Reduce method of PPrune

```
procedure REDUCE(k1, V1)
    for each r ∈ V1 do
        S[r.size] ← r
    end for
    S ← Re-index(S)
    G ← ∅
    for i = 1; i ≤ |S|; i++ do
        for each r ∈ S[i] do
            if i == 1 then
                G ← G ∪ r
            else if ! Cover(G, r) then
                G ← G ∪ r
            end if
        end for
    end for
    k2 ← k1;
    V2 ← G;
    Emit(k2, V2)
end procedure
```

4 Results and Analysis

In this section, we present the experimental results and performance analysis of the proposed algorithms. First the experimental settings, such as the computer system, the software tools, the datasets used for the experiments, and the performance measures used to evaluate the proposed algorithms are presented. Subsequently, the experimental results and performance analysis of PPrune are discussed.

4.1 Experimental Setting

In this section, the computer system, the pieces of software, the benchmark datasets, and the evaluation metrics used in the experiments are presented.

4.1.1 The Computer System and Software Tools

The proposed algorithms were written in Java and Python. The experiments were conducted on a local Hadoop 2.8.1 cluster. The cluster consisted of three physical machines and five virtual nodes. The Name Node was in one machine, and the four data nodes were in two other machines. Each two data nodes were in a machine.

4.1.2 Experimental Datasets

To study the performance of the proposed algorithms, we experimented using five benchmark datasets, namely Mushroom, Chess, T10I4D100K, Webdocs, and All-Electronics. We choose these datasets because they are frequently used in related work, and they have different characteristics. Table 1 shows the name, the number of items, the number of transactions, the average transaction size, and the number of association rules in each dataset. The first four datasets are very small; hence, we replicated them and created 16 files of sizes 1, 2, 3, and 4 GB. We also labeled each of these 16 files as shown in Table 2. For example, we will refer to the 3 GB chess dataset as D2-3 and to the 2 GB Mushroom dataset as D3-2.

4.1.3 Evaluation Measures

To measure the performance of the proposed algorithms, we used four evaluation measures, namely elapsed time, speedup, scaleup, and sizeup.

Elapsed time refers to the amount of time a process takes from the time it was submitted until the time it was completed. Speedup refers to how much faster is a parallel algorithm than the corresponding sequential algorithm and is defined by Eq. 1 as

$$\text{Speedup} = \frac{T_1}{T_p} \tag{1}$$

Table 1 Experimental datasets

Dataset	Size (KB)	Items	Transactions	Avg. items per transaction	Rules
AllElectronics	1	5	9	2.6	52
Chess	335	75	3196	37	108061
Mushroom	558	119	8124	23	111790
T10I4D100K	3928	870	100000	10	5608
Webdocs	1480	5,267,656	1,692,082	61	1,231,984

Table 2 Dataset labels

Dataset	Name	1 GB	2 GB	3 GB	4 GB
AllElectronics	D1	D1-1	D1-2	D1-3	D1-4
Chess	D2	D2-1	D2-2	D2-3	D2-4
Mushroom	D3	D3-1	D3-2	D3-3	D3-4
T10I4D100K	D4	D4-1	D4-2	D4-3	D4-4

where T_1 is the elapsed time of a task done by a single CPU, and T_p is the elapsed time of the same task done in parallel by p CPUs. Ideally, we like the speedup to be p.

Scaleup refers to the ability to keep the same elapsed time when both workload and resources increase proportionally and is defined by Eq. 2 as

$$\text{Scaleup} = \frac{T_1}{T_{n,n}} \tag{2}$$

where T_1 is the elapsed time a single CPU took to finish a workload, and $T_{n,n}$ is the amount n CPUs took to accomplish n times the workload. Ideally, we like the scaleup to be 1.

Sizeup refers to how much longer a system takes if the workload increases n times and is defined by Eq. 3 as

$$\text{Sizeup} = \frac{T_n}{T_1} \tag{3}$$

where T_1 is the elapsed time to process a workload, and T_n is the elapsed time to process n times the original workload. Ideally, we like the sizeup to be less than or equal to n.

4.2 *Performance Results and Analysis of PPrune*

The cost of a MapReduce algorithm is the sum of its I/O, CPU, and communication costs. Higher communication cost reduces the performance of a MapReduce algorithm. I/O cost scales well with the number of mappers, so it has no negative effect on performance. If the percentage of CPU cost is high, then communication cost becomes less significant giving us better performance. This subsection presents the experimental results and the performance analysis of PPrune. It was experimented with 16 different datasets and up to four nodes. The elapsed time, speedup, sizeup, and scaleup figures of PPrune are presented below.

The first set of experiments was done to study how the execution time of PPrune is affected by different datasets and different number of nodes. For this experiment, we used D1-1, D2-1, D3-1, and D4-1, and varied the nodes from 1 to 4. The results of this set of experiments are shown in Fig. 1. As expected, as the number of nodes increases, the elapsed time of PPrune decreases. Also, the elapsed time increases with the number of items in a dataset. This is because transactions from datasets with more items require more time to compare. That is why D4-1 and D3-1 required more processing time than D1-1 and D2-1.

To study the speedup, sizeup, and scaleup of PPrune, we conducted twelve sets of experiments. The first three sets were done to study the speedup, sizeup, and scaleup of PPrune using the AllElectronics dataset, D1. Figures 2, 3, and 4 show the obtained results.

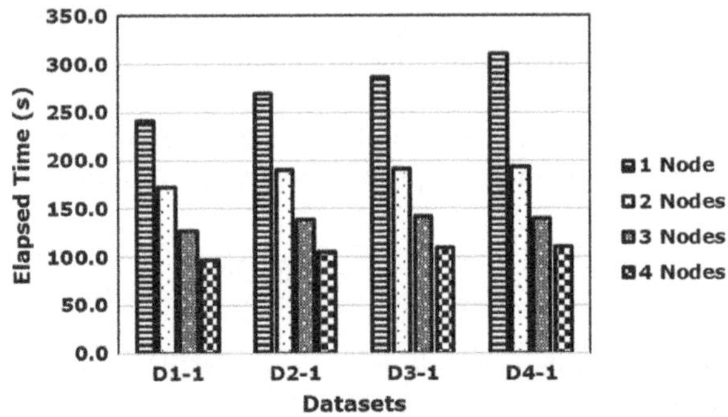

Fig. 1 PPrune elapsed time

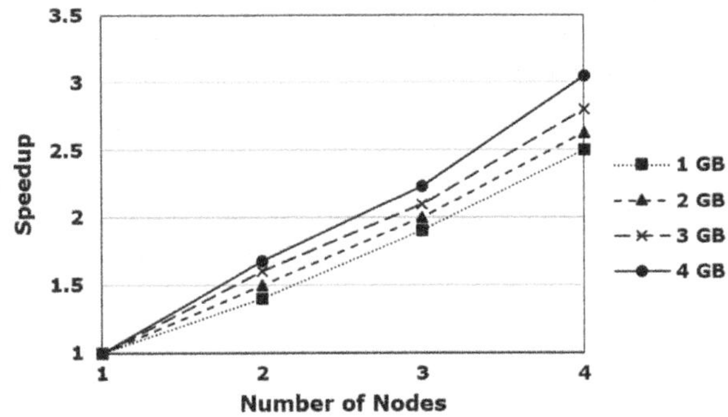

Fig. 2 Speedup of PPrune. Dataset used: D1

To measure the speedup of PPrune, we kept the size of D1 constant and varied the number of nodes from 1 to 4, Fig. 2. The speedup of PPrune improved as the size of D1 increased. With small datasets, the proportion of the communication cost is higher to that of I/O and CPU. This is because the percentage of the communication cost decreases to that of I/O and CPU costs when the dataset size increases.

To measure the sizeup of PPrune, we varied D1 size and the number of nodes as shown in Fig. 3. The results show sublinear performance, and it improved as the size of D1 increased. This is because higher percentage of time is spent on CPU and I/O than in communication.

The scaleup measure of PPrune was obtained by increasing the size of D1 in proportion with the number of nodes. Figure 4 shows the scaleup performance of the algorithm. The figure shows PPrune scales well. Scaleup decreases as the number of nodes and the size of D1 increase, but it was always more than 80%.

From the experimental results, we can easily conclude that the performance of PPrune was very good when processing dataset D1 which is characterized by few items and short transactions or rules.

To study the performance of PPrune with a dataset with more items and longer transaction than D1, we conducted three sets of experiments with D2. The results of the experiments are shown in Figs. 5, 6 and 7. The performance of PPrune was even better than with D1 because the percentage of the communication cost was higher

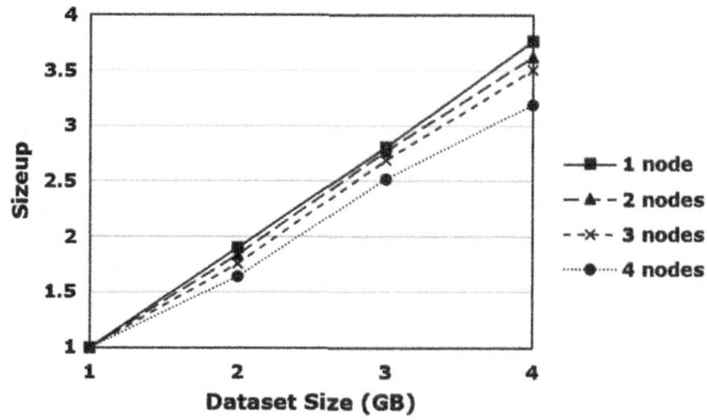

Fig. 3 Sizeup of PPrune. Dataset used: D1

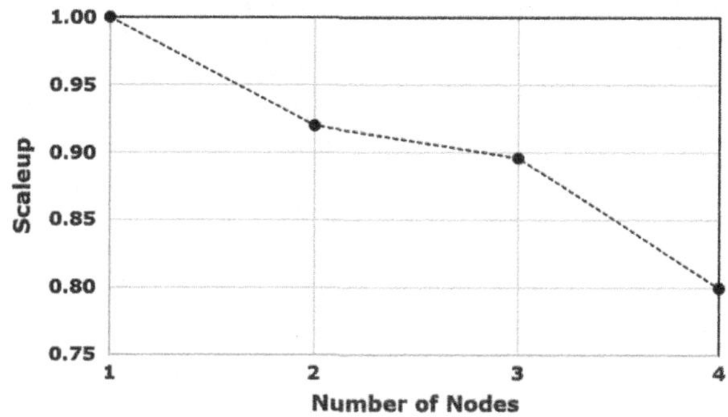

Fig. 4 Scaleup of PPrune. Dataset used: D1

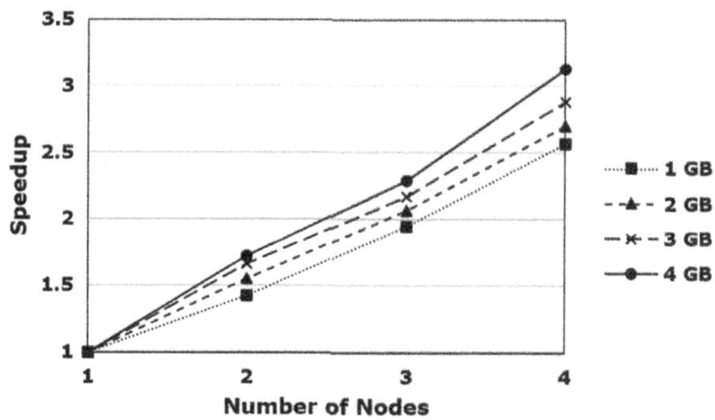

Fig. 5 Speedup of PPrune. Dataset used: D2

when processing D1 than when processing D2. Longer transaction results in longer rules which demand more CPU time than shorter rules.

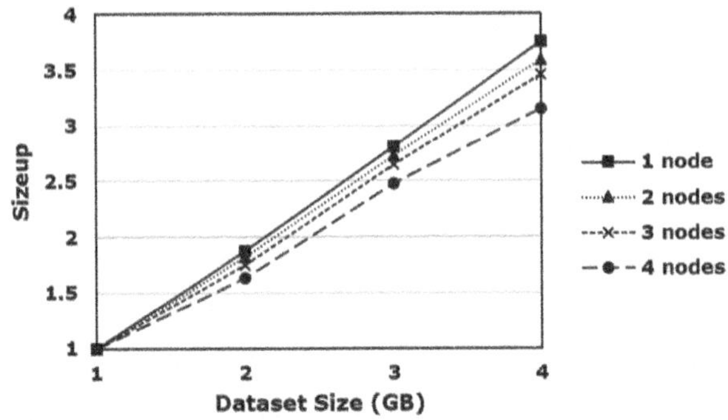

Fig. 6 Sizeup of PPrune. Dataset used: D2

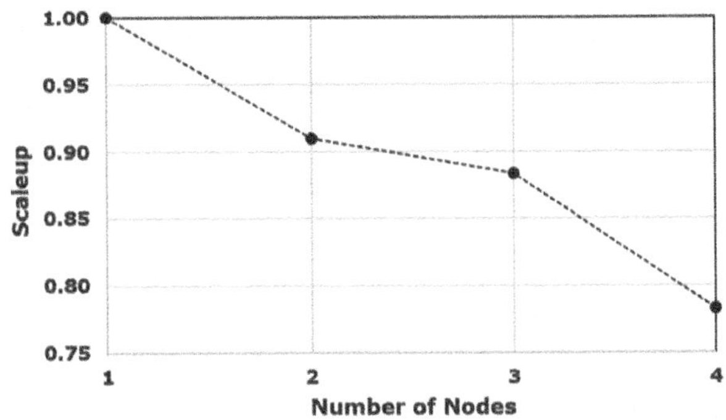

Fig. 7 Scaleup of PPrune. Dataset used: D2

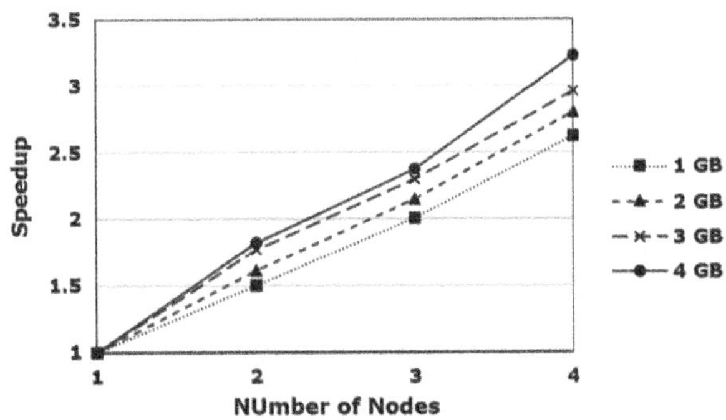

Fig. 8 Speedup of PPrune. Dataset used: D3

Dataset D3 has more items but shorter transactions than D2. The performance of PPrune when processing D3 was similar to that of D2. Having more distinct items also increases the percentage of CPU cost of PPrune. This is because it adds to the number of rules which demands longer searches. The results of the experiments are shown in Figs. 8, 9 and 10. The performance of PPrune was also better than with D1 because the percentage of the communication cost was higher when processing D1 than when processing D2.

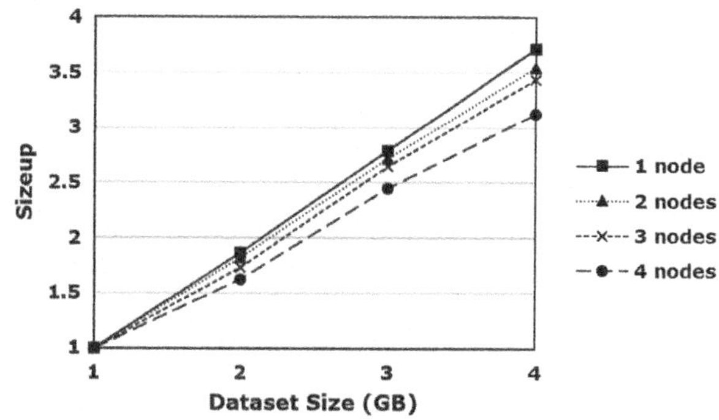

Fig. 9 Sizeup of PPrune. Dataset used: D3

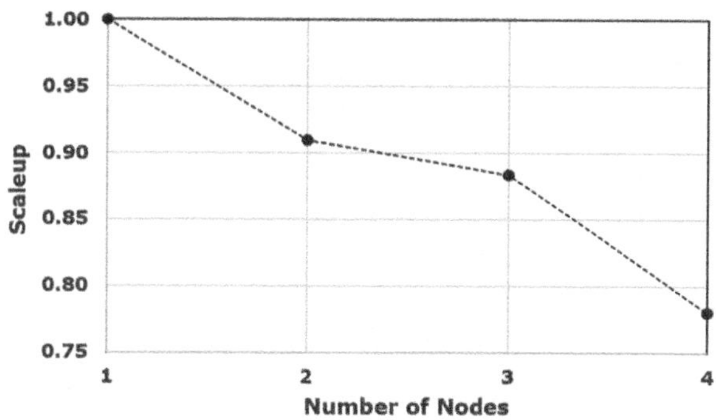

Fig. 10 Scaleup of PPrune. Dataset used: D3

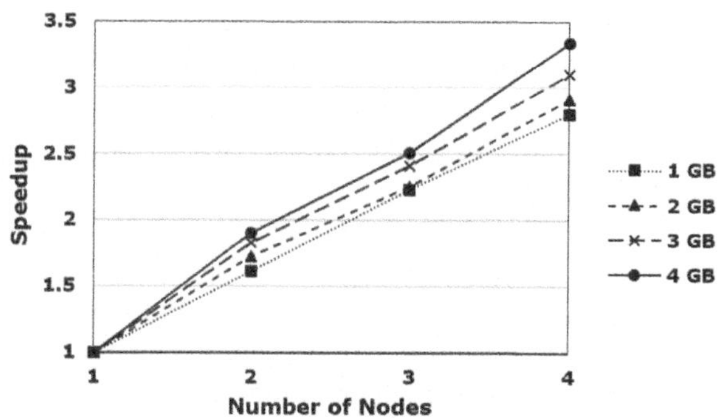

Fig. 11 Speedup of PPrune. Dataset used: D4

To study the performance of PPrune with a dataset with many number of distinct items, we conducted three sets of experiments using the dataset D4. D4 has 870 distinct items whereas D1 has only five distinct items. PPrune elapsed time is higher with D4 than with D1 as shown in Fig. 1, but the speedup, scaleup, and sizeup measures of PPrune are similar or better when processing D4. Again, more percentage of CPU time is needed to process D4 which has far more distinct items. The results of the experiments are shown in Figs. 11, 12 and 13.

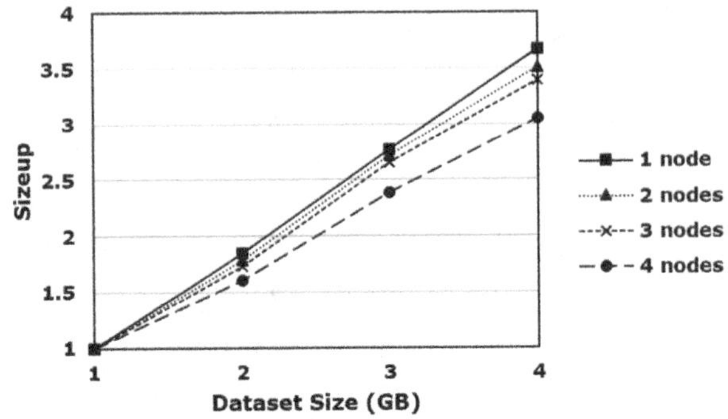

Fig. 12 Sizeup of PPrune. Dataset used: D4

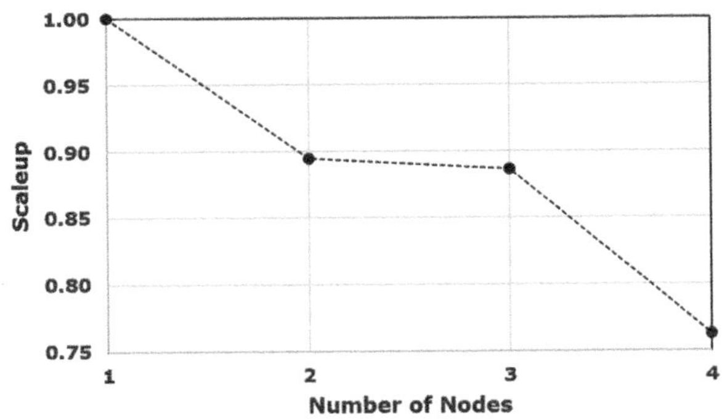

Fig. 13 Scaleup of PPrune. Dataset used: D4

5 Conclusion and Future Work

Often, the number of discovered strong association rules by ARM algorithms is large, which makes them difficult to inspect, analyze, visualize, or interpret. To solve the problem of large number of association rules, researchers proposed pruning, grouping, and clustering algorithms. Now with the advent of big data technology, the number of discovered association rules will increase drastically. Existing solutions of handling large number of discovered association rules cannot be efficiently applied to big data. To the best of our knowledge, there is no solution proposed to prune, group, or cluster rules generated from big data. This motivated us to propose, in this paper, the first solution to the huge number of association rules generated from big data. The proposed MapReduce-based algorithm, which we named PPrune, prunes association rules based on their structure. To study the performance of PPrune, we conducted many experiments on a Hadoop cluster. In the experiments, two real benchmark datasets and two synthetic benchmark datasets were used. The performance of PPrune was evaluated using four measures, namely elapsed time, scaleup, sizeup, and speedup. PPrune showed high performance. For example, the lowest scalability obtained by the experiments was 77%. As future work, the proposed work can be extended to test the proposed algorithm using more nodes and bigger datasets. We plan to use a visualization software to see the distribution of clusters and incorporate other pruning techniques. We also work on extending the proposed solution to handle rules with unlimited number of items in their consequents.

Acknowledgements The authors would like to thank King Fahd University of Petroleum and Minerals (KFUPM), Saudi Arabia, for the support during this work.

References

1. Agrawal R, Imieliński T, Swami A (1993) Mining association rules between sets of items in large databases. In: ACM sigmod record, vol 22. ACM, pp 207–216
2. Aijun A, Khan SM, Huang X (2003) Objective and subjective algorithms for grouping association rules. In: Proceedings of 3rd IEEE international conference on data mining (ICDM'03), vol 3, p 477
3. Bay SD, Pazzani MJ (2001) Detecting group differences: mining contrast sets. Data Mining Knowl Discov 5(3):213–246
4. Bayardo RJ, Agrawal R, Gunopulos D (2000) Constraint-based rule mining in large, dense databases. Data Mining Knowl Discov 4(2–3):217–240
5. Berrado A, Runger GC (2007) Using metarules to organize and group discovered association rules. Data Mining Knowl Discov 14(3):409–431
6. Brijs T, Vanhoof K, Wets G (2000) Reducing redundancy in characteristic rule discovery by using IP-techniques. Intell Data Anal J 4:200
7. Chawla S, Davis J (2003) On local pruning of association rules using directed hypergraphs. In: Proceedings of 20th IEEE international conference on data engineering (ICDE'04)
8. Chawla S, Davis JG, Pandey G (2004) On local pruning of association rules using directed hypergraphs. In: ICDE, vol 4, pp 832–841
9. Chen MS, Han J, Yu PS (1996) Data mining: an overview from a database perspective. IEEE Trans Knowl Data Eng 8(6):866–883
10. Hahsler M, Karpienko R (2017) Visualizing association rules in hierarchical groups. J Bus Econ 87(3):317–335
11. Han J, Pei J, Yin Y, Mao R (2004) Mining frequent patterns without candidate generation: a frequent-pattern tree approach. Data Mining Knowl Discov 8(1):53–87
12. Huang S, Webb GI (2005) Discarding insignificant rules during impact rule discovery in large, dense databases. In: Proceedings of SIAM international conference on data mining, pp 541–545
13. Lent B, Swami A, Widom J (1997) Clustering association rules. In: Proceedings of 13th international conference on data engineering, pp 220–231
14. Liu B, Hsu W, Ma Y (1998) Integrating classification and association rule mining. In: Proceedings of 4th international conference on knowledge discovery and data mining
15. Liu B, Hsu W, Ma Y (1999) Pruning and summarizing the discovered associations. In: Proceedings of the fifth ACM SIGKDD international conference on knowledge discovery and data mining. ACM, pp 125–134
16. Marinica C, Guillet F, Briand H (2008) Post-processing of discovered association rules using ontologies. In: IEEE International Conference on data mining workshops, 2008. ICDMW'08. IEEE, pp 126–133 (2008)
17. Martín FJV, Sequera JLC, Huerga MAN (2017) Using data mining techniques to discover patterns in an airline's flight hours assignments. Int J Data Warehous Mining (IJDWM) 13(2):45–62
18. Strehl A, Gupta GK, Ghosh J (1999) Distance based clustering of association rules. In: Proceedings ANNIE, vol. 9, pp 759–764

19. Toivonen H, Klemettinen M, Ronkainen P, Hatonen K, Mannila H (1995) Pruning and grouping discovered association rules. In: ECML'95 MLnet workshop on statistics, machine learning, and knowledge discovery in databases, pp 47–52
20. Yang L (2003) Visualizing frequent itemsets, association rules, and sequential patterns in parallel coordinates. In: International conference on computational science and its applications. Springer, pp 21–30

A Statistical Analysis on KDD Cup'99 Dataset for the Network Intrusion Detection System

Satish Kumar, Sunanda, and Sakshi Arora

Abstract Knowledge Discovery and Data Mining Tools Competition (KDD Cup'99) has been used on prior choice by researchers for the purpose of research and simulation for network intrusion detection system (NIDS) as well as intrusion prevention system (IPS) in network security. In this paper, a statistical analysis on the KDD Cup'99 dataset has been done for the objective of better understanding of structural and organizational view of the features/attributes of network traffic data. This analysis is required for evaluating various predictions on the attributes of the network traffic by taking consideration of different statistical aspects of KDD Cup'99 dataset.

Keywords Dataset · Features · Intrusion detection system · KDD Cup'99 dataset · Statistical significance

1 Introduction

Today is an era of networking and Internet. A number of hosts/terminals are continuously increasing in this scenario. So, possibility of breaching information and vulnerability to security is increasing tremendously.

Intrusion detection system (IDS) is a device or software that monitors malicious activities in a network or a computer system [4, 5]. These IDSs can be divided into three categories based on the installation in a system—host-based IDS (HIDS), network-based IDS (NIDS) and specification-based IDS. The basic goal of an IDS is to identify an intrusion, logging information and reporting about that incidents to network administrator. Various datasets such as DARPA98, KDD99, NSL-KDD'99, ISC2012, ADFA13 and UNSW-NB15 [19] have been used by researchers to evaluate

S. Kumar (✉)
School of CSE, SMVD University, Katra, Jammu 182320, India
e-mail: 16dcs005@smvdu.ac.in

Sunanda · S. Arora
Faculty of Engineering, School of CSE, SMVD University, Katra, Jammu 182320, India
e-mail: sunanda.gupta@smvdu.ac.in

S. M. Thampi et al. (eds.), *Applied Soft Computing and Communication Networks*, Lecture Notes in Networks and Systems 125,
https://doi.org/10.1007/978-981-15-3852-0_9

the performance of applied approaches in intrusion detection system [23].Mahbod et al. [17] reveal that machine learning has been used for detection rate up to 98% with a false-positive rate less than 1%.

Dataset KDD Cup'99 being a reliable IDS benchmark is a labeled intrusion detection dataset. University of California, Irvine (UCI) provides a publically free dataset named KDD Cup'99, International Knowledge Discovery and Data Mining Tools Competition, for researchers and designers of IDS who use KDD Cup'99 as a benchmark on which different methodologies are evaluated.

The overall aim of this paper is to analyze that how the KDD Cup'99 dataset is distributed and organized. Besides this, the aims of statistical analysis in this paper are

- To know the structure and pattern of the KDD Cup'99 dataset which has been used as a benchmark dataset for network intrusion detection system.
- To measure central tendency of features around the mean of features.
- To study the interrelationship among attributes of the KDD Cup'99 dataset.
- To measure the association between the features of the benchmark dataset.
- To study the association between columns and rows of the benchmark dataset and behavior of network traffic data.

The other sections of this paper have been organized as follows: Sect. 2 explores definition, history and structural organization of KDD Cup'99. Section 3 tells about the used methodology in this work on KDD'99 datasets. Section 4 evaluates contribution of previous work done and related articles with common mistakes and strength points in the field of NIDS and KDD Cup'99. Section 5 gives descriptive statistics along with various statistical methods like mean value significance, standard deviation, linear correlation coefficient (LCC) and analysis of variance (ANOVA) about KDD Cup'99. Section 6 suggests about experimental environment with required hardware and software used. Finally, Sect. 7 suggests further studies in order to improve the quality of similar studies and discusses the results of this work done.

2 Related Work

Denning [5] proposed an independent model of any particular system or system vulnerability of any type of intrusion. This paper also reveals the statistical modeling based on the assumption of mean and standard deviation that if a new observation $Xi + 1$ is observed and it assumed that it is not normal traffic record, then it will fall beyond the 'confidence interval' which is based on the mean and standard deviation of a feature/attribute in a dataset.

Lippmann et al. [16] made understandable attack mechanism, analyzed the task of verifying each alert, determining attacker's actions and responding to an attack which is helpful to reduce the susceptibility of same types of attacks and revealed enduring valuable records of attack-related events. A detailed successful evaluation of 58 types of attacks on DARPA 1999 was performed by Lippmann et al. [16].

DARPA 1999 was an offline dataset, and this was the initial step to measure false alarm rate and different attack types in intrusion detection system. This work also revealed the evolution of KDD Cup'99 dataset, both real time and offline.

McHugh [18] identified the basic characteristics along with some problems of the Lincoln Laboratory effort that arose due to the traffic data recorded from DARPA'98. This article focused on DARPA 1998 evaluation with brief discussion on KDD Cup'99 dataset evaluation. Both 1999 and 1998 evaluations demonstrated that research systems were very poor for detecting new attacks. This article also described the risk regarding systems that may be biased toward unrealistic expectation with respect to true detection, false alarm or both.

Mahbod et al. [17] proposed NSL-KDD with the drawbacks of KDD Cup'99 dataset. This paper also enlightened the probability distribution of network traffic records and their behavior in the training and test set data. It revealed that some specific attack types are not included in the training data but are included in the test set. So, use of KDD Cup'99 makes the task of intrusion detection more practical. This paper also focused on the statistics of randomly selected traffic records, their classification of features, attack categories with space, record counts along with various versions of KDD Cup'99 as KDD train + and KDD test + dataset.

Winters, Winters and Amedee [26] explored some selected important concepts such as parametric and nonparametric test, mean, Gaussian distribution, correlation, standard deviation, T-test, Z-test and F-test. These basic concepts have been used in the analysis of data.

Kavousi and Akbari's [10] paper helps to find out the identification of the attack behavior patterns. This paper revealed the attack strategies through automatic analysis of intrusion alerts. A new algorithm to mine attack behavior patterns from a large number of intrusion alerts without specific prior knowledge about attacks was discussed by the author. A Bayesian mechanism for automatic generation of correlation rules from previously observed alerts is presented in this paper. This new introduced approach helps us to predict forthcoming attack in a real-time system.

Özgür and Erdem [22] described the history and organization of KDD Cup'99. It has also been revealed that the KDD Cup'99 has been used in many studies during last 16 years along with the citation in many studies. Moreover, 149 research articles had used KDD Cup'99 those were published in Science Citation Index Expanded (SCIE) and Emerging Sources Citation Index (ESCI) journals between 2010 and 2015.

Reference [2] gave a brief outline of basic statistical methods used to conduct a research study on variables and provided an understanding of quantitative and qualitative variables along with the measures of central tendency. It gives a brief idea of the sample size estimation, power analysis and the statistical errors along with a summary of parametric and nonparametric tests that are used for data analysis.

In the previous literature, no such statistical analysis of a benchmark dataset has been found that presents statistical measures for central tendency, association among features, interrelationship between features, interrelationship between rows and columns of the dataset. So, there is a strong need to analyze a benchmark dataset on statistical basis.

3 KDD Cup'99 and History

In 1998 and 1999, the Lincoln Laboratory under the sponsorship of Defense Advanced Research Projects Agency (DARPA) and Air Force Research Laboratory (AFRL) at Massachusetts Institute of Technology generated standard network traffic data to review and access research activities in the field of network intrusion detection [13]. An environment was set up to acquire a TCP/IP dump in a LAN with multiple attacks having 4,898,431 traffic data records out of which 3,925,650 were attacks only [8]. Further, a special interested group of Association for Computing Machinery (ACM) named Knowledge Discovery and Data Mining (KDD) organized the annual data mining and discovery competition for year 1999 called KDD Cup'99 focused on computer network intrusion detection area; this KDD Cup'99 dataset is freely available for researchers and designers of IDS. This is a benchmark dataset on which different methodologies are evaluated.

The overall KDD Cup'99 dataset has been categorized into three basic components of the KDD Cup'99 dataset Lippmann, et al. [16].

1. 10% KDD Labeled Training Dataset—This part of KDD Cup'99 is considered as training data and contains 97278 normal records out of total 494021 records.
2. KDD Cup'99 Test Data—This portion of the KDD Cup'99 has been considered as test dataset which further modified with less redundant traffic data packets and called as Corrected KDD with 60593 normal records among total of 311029 records.
3. Whole KDD—This is the whole KDD Cup'99 dataset having 972780 normal records out of 4898431 records.

This dataset contains 5 millions of records with labeled training data, labeled test data and unlabeled whole data. Each traffic data has 41 features. These 41 features are shown in Table 1, and these features are categorized into three groups as [7, 12, 17]:

a. Basic Features: These features are derived from packet headers without inspecting payload.
b. Content Features: In this feature, domain knowledge is used to access the payload of the original TCP packets.
c. Time-Based Traffic Features: These are the features that are used to capture properties that mature over a two-second temporal window.

The overall feature in KDD Cup'99 dataset [11, 15] has been categorized into continuous and discrete type also. All these continuous or discrete types of features are categorized in Table 1.

The network traffic data that is identified as 'bad' connection and vulnerable to the security to the system is called attacks, and this type of traffic data is not considered as 'normal' type. Besides the increasing issue of anomaly, unknown attacks and the basic network attacks [9, 11], KDD Cup'99 attacks have been classified into four categories, viz. DoS, R2L, U2L and probe. All the traffic records of attack types fall in one of the following four categories [17, 21].

Table 1 Overall features with label, types and classes in KDD Cup'99 dataset

Feature name with label	Type	Description	Feature name with label	Type	Description
Duration1	Continue/basic	Number of seconds of the connection	IsGust_LogIn22	Discrete/content	1—The login belongs to the 'Guest' list 0—Otherwise
Proto_Type2	Discrete/basic	Type of the protocol	Host_Count23	Continue/traffic	Number of connections to the same host as the current connection in the past two seconds
Service3	Discrete/basic	Network service on the destination	Srv_Count24	Continue/traffic	Number of connections to the same service as the current connection in the past two seconds
Flag4	Discrete/basic	Normal or error status of the connection	Syn_ErrR25	Continue/traffic	% of connections that have 'SYN' errors to same-host connections
Src_Byte5	Continue/basic	Number of data bytes from source to destination	SrvSyn_ErrR26	Continue/traffic	% of connections that have same-server 'SYN' errors
Dst_Byte6	Continue/basic	Number of data bytes from destination to source	Rej_ErrR27	Continue/traffic	% of connections that have 'REJ' errors

(continued)

Table 1 (continued)

Feature name with label	Type	Description	Feature name with label	Type	Description
Land7	Discrete/basic	1—Connection is from/to the same host/port 0—Otherwise	SrvRej_ErrR28	Continue/traffic	% of connections that have 'REJ' errors to same-service connections
Wrng_Frgment8	Continue/basic	Number of 'wrong' fragments	Same_SrvR29	Continue/traffic	% of connections that have same-service connections
Urgnt9	Continue/basic	Number of urgent packets	Diff_SrvR30	Continue/traffic	% of connections to different services to same-host connections
Hot10	Continue/content	Count of access to system directories, creation and execution of programs	Srv_diffHostR31	Continue/traffic	% of connections to different hosts to same-service connections
Failed_LogIn11	Continue/content	Number of failed login attempts	Dst_HCount32	Continue/traffic	Host_Count23 for destination host
Logged_In12	Discrete/content	1—Successful login 0—Otherwise	Dst_HsrvCount33	Continue/traffic	Srv_Count24 for destination host
Num_Compro13	Continue/content	Number of 'compromised' condition	SmDstH_SmSrvc34	Continue/traffic	Same_SrvR29 for destination host
Root_Shell14	Discrete/content	1—Root shell is obtained 0—Otherwise	DstH_DiffSrvR35	Continue/traffic	Diff_SrvR30 for destination host

(continued)

Table 1 (continued)

Feature name with label	Type	Description	Feature name with label	Type	Description
su_Attempt15	Discrete/content	1—'su root' command attempt 0—Otherwise	DstH_SmSrcPortR36	Continue/traffic	Same source port rate for destination host
Num_Root16	Continue/content	Number of 'root' access	DstHSrvc_DiffHstR37	Continue/traffic	% of connections to different hosts for destination host
Num_FCreations17	Continue/content	Number of file creation operations	DstHst_SErrR38	Continue/traffic	% of connections that have 'SYN' errors to destination host
Num_ShellPrmt18	Continue/content	Number of shell prompts	DstHstSrvc_SErrR39	Continue/traffic	SrvSyn_ErrR26 for destination host
Num_Facc19	Continue/content	Number of operations on access control files	DstHst_RSTerrR40	Continue/traffic	Rej_ErrR27 for destination host
Num_OBndCMD20	Continue/content	Number of outbound commands in an ftp session	DstHstSrvc_RSTer41	Continue/traffic	SrvRej_ErrR28 for destination host
Is_HotLogIn21	Discrete/content	1—The login belongs to the 'hot' list 0—Otherwise	Behaviour42	–	Attack types

1. Denial of Services (DoS): Attacker/s make/s flooding of superfluous request on the target machine and hence keep/s busy the memory and computing resources of the machine occupied so as to avoid the fulfillment of legitimate request of users to the machine. Examples are apache2, back, land, mailbomb, SYN Flood, smurf, syslogd, teardrop, udpstorm, etc.
2. User to Root Attack (U2R): Attacker/s begin/s by accessing as normal user account and become/s capable to exploit vulnerability to root access in the system. Examples for these types of attacks are Eject, Ffbconfig, Fdformat, loadmodule, perl, ps and Xterm.
3. Remote to Local Attack (R2L): This class of attack is also known as user to super-user. Attacker/s send/s packets to a machine on which it does not have account as a user and exploit/s vulnerability to gain local user accessibility on that machine. Examples are Dictionary, ftp_write, imap, Guest, phf, named, sendmail, xlock and xsnoop.
4. Probe: Attacker/s scan/s a network to collect information about computers in that network for the purpose of circumventing its security controls. Examples are ipsweep, Mscan, Nmap, Saint and Satan.

According to [8, 14, 21, 24, 25], all types of attacks with their count and percentages of test as well as training set of KDD Cup'99 are described in Tables 2 and 4. The dataset contains a total of 24 training attack types with an addition of 14 types of test data. Here, it is important to point out that the test set is not from the same probability distribution as the training set. It is also to be noted that test set has included specific attack types that are not in the training data. Some attacks are only present in test set but not present in training set and vice versa. Furthermore, HTTPtunnel attack is described as U2R class of attack in [6], but [16] included it in R2L class. In this paper also, HTTPtunnel has been included in R2L class of attacks.

KDD'99 is based on the organization of traffic data recorded in DARPA'98. McHugh [18] described the main disadvantage which is characterized due to the synthetic data as in the DARPA'98. As a result, some of the existing problems as in DARPA'98 also remained in KDD'99. The most important deficiency according to [18] in the KDD dataset is that there is a huge number of redundant records and on analyzing KDD training and test sets, it has been found that about 78 and 75% of the records are duplicated in the training and test sets, respectively. Such a large amount of redundant records in the training set will cause learning algorithms to be biased toward the more frequent redundant records. On the one hand, the existence of repeated redundant records in the test set will cause the evaluated result to be biased toward the methods which have better detection rates on the frequent records (Fig. 1).

Table 2 Various types of attacks described in test data in KDD Cup'99

Attack	Types	Functioning	Impact
apache2	DoS	Abuse	Slow down server response/crash down httpd
back	DoS	Abuse/bug	Slow down server response
buffer_overflow	U2R	Abuse	Gains root shell
ftp_write	R2L	Misconfiguration	Gain user access
guess_passwd	R2L	Login misconfiguration	Gain user access
HTTPtunnel	R2L	Misconfiguration	Gain local access from a remote machine
imap	R2L	Bug	Gain user access
ipsweep	Probe	Abuse of feature	Identifies active machine
land	DoS	Bug	low down server response
loadmodule	U2R	Poor environment sanitation	Gains root shell
mailbomb	DoS	Abuse	Overflowing server mail queue/annoyance
Mscan	Probe	Abuse of feature	Looks for known vulnerability
multihop	R2L	Misconfiguration	Multi-day scenario in which a user first breaks into one machine
named	R2L	Bug	Gains root shell
Neptune	DoS	Abuse	Slow down server response
Nmap	Probe	Abuse of feature	Identifies active ports on a machine
normal	Normal	Normal	Good connection
perl	U2R	Poor environment sanitation	Gain root shell
phf	R2L	Bug	Executes commands as root
pod	DoS	Abuse	Slow down server response
portsweep	Probe	Abuse of feature	Identifies active ports on a machine
Processtable	DoS	Abuse	Deny new processes
ps	U2R	Poor temp file management	Gain root shell
rootkit	U2R	Enable privileged access	Multi-day scenario where a user installs one or more components of a rootkit
Saint	Probe	Abuse of feature	Looks for known vulnerabilities
Satan	Probe	Abuse of feature	Looks for known vulnerabilities
sendmail	R2U	Bug	Executes commands as root
smurf	DoS	Abuse	Slows down the network

(continued)

Table 2 (continued)

Attack	Types	Functioning	Impact
snmpgetattack	R2L	Misconfiguration	Gain access to system using SNMP services.
snmpguess	R2L	Login misconfiguration	Attempts to login by guessing the password using SNMP service.
Sqlattack	U2R	Abuse of Web application's database	To gain access to the database of the Web and its application.
Teardrop	DoS	Bug	Reboots the machine
udpstorm	DoS	Abuse	Slows down the network
Warezmaster	R2L	Abuse	Gain user access
worm	R2L	Misconfiguration	Exploit other hosts through network
xlock	R2L	Misconfiguration	Spoof user to access password
xsnoop	R2L	Misconfiguration	Monitor keystrokes remotely
Xterm	U2R	Buffer overflow	Gain root shell

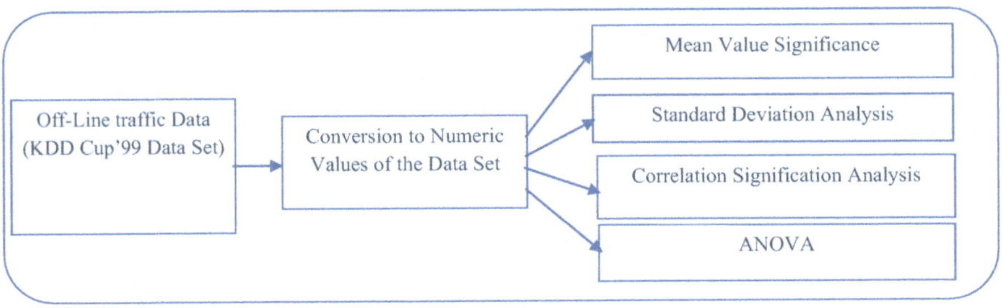

Fig. 1 Statistical analysis methodology for KDD Cup'99 dataset

4 Proposed Methodology

Since the used dataset, KDD Cup'99 is large in scale. To analyse the statistical values (Mean, Standard Deviation, Correlation Co-efficient and ANOVA) of the test set in the KDD Cup'99 dataset, the dataset is partitioned into normal and intrusive/attacked dataset. Then, compare the statistical value of features related to normal traffic data with the features related to intrusive traffic data. The proposed analysis is based on the resemblance, co-relation, dependency and significance among features of normal and intrusive traffic data.

5 Experimental Environment

KDD Cup'99 (Knowledge Discovery and Data Mining) is the popular dataset, publically available for research and design. It can be freely downloaded from the University of California Irvine Machine Learning Repository on URL http://archive.ics.uci.edu/ml/datasets/KDD+Cup+1999+Data. The computer hardware used in this analysis is Intel Core^{TM} i3, 2.1 GHz CPU, 4 GB memory, 500 GB hard disk with software environment Windows 7, 64-bit operating system and MATLAB R2017b.

6 Statistical Analysis of KDD Cup'99 Dataset

Statistical analysis explores the relation and nature of the data to be analyzed. This analysis is also helpful to find patterns in structured and semi-structured dataset for fast detection rate in decision engine [20] with less numbers of false-positive detection. This study on test set of KDD Cup'99 will be helpful in summarizing understanding of how the data relates to the underlying dataset.

Features like Proto_Type2, Service3 and Flag4 are of character-type data. Hence, the training and test sets of the KDD Cup'99 dataset, first of all, are to be converted into numeric values so that the value of these character-type features is assigned to some numeric values for the sake of statistical calculation.

a. *KDD Cup'99 Test and Training Dataset*

Table 3 explores that 'normal' traffic data has a counting of 97278 with 19.69% records while 'attack' traffic data has a counting of 396743 with 80.31% records out of a total 494021 network traffic records in the training set of KDD Cup'99. In the test set of KDD Cup'99, there is a 60593 of 'normal' traffic data with 19.48% records while 'attack' has a counting of 250436 with 80.52% records out of a total 311029 network traffic records.

The types of attack with their count and percentages of KDD Cup'99 test as well as training data have been given in Table 4. This table shows 'normal' and all type of 'attacks' in traffic record data with their percentage in both training and test datasets of KDD Cup'99 dataset. Some attacks are only present in test set but not present in training set and vice versa.

Table 3 Count and percentage of normal as well as attack traffic records in test and training dataset of KDD Cup'99

Value	Count	Percent	Count	Percent
	Test (%)		Training (%)	
Normal	60593	19.48	97278	19.69
Attack	250436	80.52	396743	80.31

Table 4 Types of attack behavior with their percentage count in both training and test sets of KDD Cup'99 dataset

Value	Count	Percent	Count	Percent	Value	Count	Percent	Count	Percent
	Test (%)		Training			Test (%)		Training	
apache2	794	0.26	–	–	portsweep	354	0.11	1040	0.210517367
back	**1098**	**0.35**	**2203**	**0.44593246**	Processtable	**759**	**0.24**	**–**	**–**
buffer_overflow	22	0.01	30	0.006072616	ps	16	0.01	–	–
ftp_write	**3**	**0.00**	**8**	**0.001619364**	rootkit	**13**	**0.00**	**10**	**0.002024205**
guess_passwd	4367	1.40	53	0.010728289	Saint	736	0.24	–	–
HTTPtunnel	**158**	**0.05**	**–**	**–**	Satan	**1633**	**0.53**	**1589**	**0.321646246**
imap	1	0.00	12	0.002429047	sendmail	17	0.01	–	–
ipsweep	**306**	**0.10**	**1247**	**0.252418419**	smurf	**164091**	**52.76**	**280790**	**56.8376648**
land	9	0.00	21	0.004250831	snmpgetattack	7741	2.49	–	–
loadmodule	**2**	**0.00**	**9**	**0.001821785**	snmpguess	**2406**	**0.77**	**–**	**–**
mailbomb	5000	1.61	–	–	Spy	–	–	2	0.000404841
Mscan	**1053**	**0.34**	**–**	**–**	Sqlattack	**2**	**0.00**	**–**	**–**
multihop	18	0.01	7	0.001416944	Teardrop	12	0.00	979	0.198169713
named	**17**	**0.01**	**–**	**–**	udpstorm	**2**	**0.00**	**–**	**–**
Neptune	58001	18.65	107201	21.69968483	warezclient	–	–	1020	0.206468956
Nmap	**84**	**0.03**	**231**	**0.046759146**	Warezmaster	**1602**	**0.52**	**20**	**0.004048411**
normal	60593	19.48	97278	19.69106576	worm	2	0.00	–	–
perl	**2**	**0.00**	**3**	**0.000607262**	xlock	**9**	**0.00**	**–**	**–**
phf	2	0.00	4	0.000809682	xsnoop	4	0.00	–	–
pod	**87**	**0.03**	**264**	**0.053439024**	Xterm	**13**	**0.00**	**–**	**–**

This subsection presents the overall structure and pattern of the KDD Cup'99 dataset with the count and percentage of each network traffic record.

b. *Mean Value Significance (MVS)*

The mean values of the attributes in KDD Cup'99 have been used to analyze central tendency. If N is the total number of items of attributes v_i in a dataset, then mean $\bar{x}$ will be as given below

$$\bar{x} = \frac{1}{N}\sum_{i=1}^{N} v_i \tag{1}$$

The mean values of the test set in the KDD Cup'99 shown in Table 5 explain that the mean values of those features whose value is of zero or very near to zero have low relevancy in the attack detection. In the analysis of traffic records, it has been found that those features whose mean values are zero or near to zero have less significant variations in the values of features and they have central tendency to zero. It means

Table 5 Features with their mean values in test set of KDD Cup'99 dataset

Feature's name	Mean values (training set)	Mean values (test set)	Feature's name	Mean values (training set)	Mean values (test set)
Duration1	47.9793	17.90273576	IsGust_LogIn22	0.001387	0.002424211
Proto_Type2	1.467132	1.555456244	Host_Count23	332.2857	269.2470188
Service3	24.40889	26.16030981	Srv_Count24	292.9066	235.5800392
Flag4	8.842446	8.642901466	Syn_ErrR25	0.176687	0.059215282
Src_Byte5	3025.61	1731.702192	SrvSyn_ErrR26	0.176609	0.059193226
Dst_Byte6	868.5324	747.9936855	Rej_ErrR27	0.057433	0.142585257
Land7	4.45E-05	2.89E-05	SrvRej_ErrR28	0.057719	0.142247636
Wrng_Frgment8	0.006433	0.000761987	Same_SrvR29	0.791547	0.815653524
Urgnt9	1.42E-05	5.14E-05	Diff_SrvR30	0.020982	0.024446563
Hot10	0.034519	0.014677088	Srv_diffHostR31	0.028997	0.025349051
Failed_LogIn11	0.000152	0.002363124	Dst_HCount32	232.4708	235.282681
Logged_In12	0.148247	0.172475878	Dst_HsrvCount33	188.6657	199.1939144
Num_Compro13	0.010212	0.011243325	SmDstH_SmSrvc34	0.75378	0.793494208
Root_Shell14	0.000111	0.000199338	DstH_DiffSrvR35	0.030906	0.024952625
su_Attempt15	3.64E-05	2.25E-05	DstH_SmSrcPortR36	0.601935	0.547918586
Num_Root16	0.011352	0.008359349	DstHSrvc_DiffHstR37	0.006684	0.00456623
Num_FCreations17	0.001083	0.00095811	DstHst_SErrR38	0.176754	0.058764199
Num_ShellPrmt18	0.000109	8.36E-05	DstHstSrvc_SErrR39	0.176443	0.058791174
Num_Facc19	0.001008	0.000771632	DstHst_RSTerrR40	0.058118	0.142658852
Num_OBndCMD20	0	0	DstHstSrvc_RSTerrR41	0.057412	0.141693347
Is_HotLogIn21	0	3.86E-05	Behviour42	15.54627	22.61246057

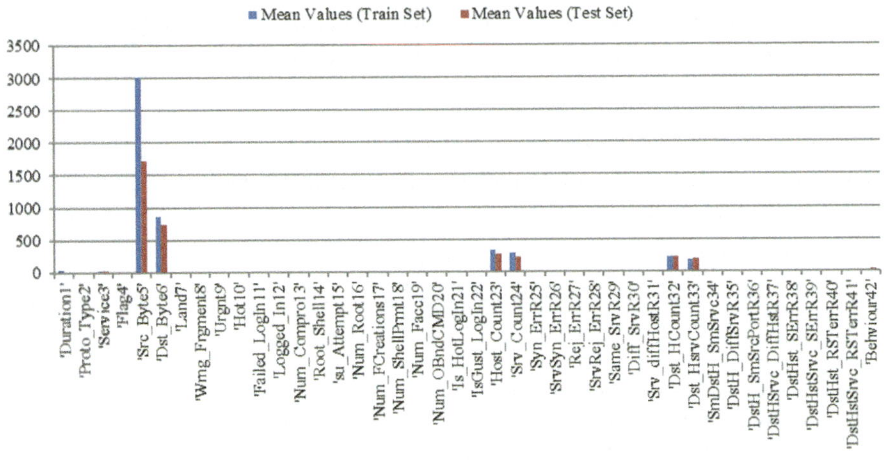

Fig. 2 Mean of various features of training and test sets in KDD Cup'99

that there is either no outlier or they are very less in numbers having low values. In Table 5, features 1-6, 10, 12, 23, 24, 25, 26, 27, 28, 29, 30, 31, 32-34, 35, 36, 38, 39, 40, a total of 25 features show statistical significance of features in the test set of KDD Cup'99 dataset.

Figure 2 shows the visualization of feature's mean value for test dataset and also explores the man statistical significance of the features in the test set of KDD Cup'99.

Table 6 explains the mean values of every feature for each type of behavior of the traffic data type. Mean value for the same feature of different behaviors also shows that how the data of same features is spread out which has zero or very near value to zero. On the other hand, if a new observation, say v_i, is to be assumed that not of normal behavior then it has been considered that it will fall outside of the 'confidence interval' [5]. For example, apache2 type of behavior in the test set has the statistical significance with 22 features while buffer_overflow attack has the statistical significance for 28 different features in the test set.

Features 8 and 9 (Wrng_Frgment8 and Urgnt9, respectively) to the behavior land, mailbomb, Mscan, multihop, named, Neptune and Nmap in Table 6 have all zero values, which means that spread of these features' data, feature labels 8 and 9, remains with zero values, while feature 6 (Dst_Byte6) has a wide range of data spread for behavior named as land, mailbomb, Mscan, multihop, named, Neptune and Nmap in the same table. So, a different behavior of the traffic data has a wide range of the data values for the same features and there is a large variety for the behavior of the traffic data with same feature.

Maximum and minimum values in Table 7 describe that there is huge differences between the values of corresponding traffic record/behavior of the same type. So, requirement of normalization of the traffic data in the KDD Cup'99 in a network intrusion detection system is needed.

Table 6 Mean values of features for behavior of traffic records

Feature name	apache2	buffer_overflow	ftp_write	guess_passwd	HTTPtunnel	imap	ipsweep
Duration1	**785.1410579**	**300.6363636**	**15.333333**	**3.485459125**	**18.74050633**	91	**0**
Proto_Type2	2	2	2	2	2	2	1
Service3	**23**	**48.54545455**	**23.666667**	**46.12205175**	**41.81012658**	**26**	**14.01960784**
Flag4	5.400503778	10	10	9.951912068	3.202531646	3	10
Src_Byte5	**39030.97229**	**1788.318182**	**54.666667**	**42.9187085**	**420.664557**	**1352**	**26.30718954**
Dst_Byte6	1.037783375	19085.36364	1108.6667	112.8264255	340.3291139	291	0
Land7	**0**	**0**	**0**	**0**	**0**	**0**	**0**
Wrng_Frgment8	0	0	0	0	0	0	0
Urgnt9	**0**	**0**	**0.3333333**	**0**	**0**	**0**	**0**
Hot10	0	3.318181818	0.3333333	0	0.025316456	0	0
Failed_LogIn11	**0**	**0.090909091**	**0**	**0.162583009**	**0**	**0**	**0**
Logged_In12	0.654911839	0.818181818	0.6666667	0.83650103	0.012658228	0	0
Num_Compro13	**0**	**6.863636364**	**0**	**0**	**0.056962025**	**0**	**0**
Root_Shell14	0	0.590909091	0	0	0.006329114	0	0
su_Attempt15	**0**	**0**	**0**	**0**	**0.006329114**	**0**	**0**
Num_Root16	0	4.272727273	0	0	0.044303797	0	0
Num_FCreations17	**0**	**1.318181818**	**0.3333333**	**0**	**0**	**0**	**0**
Num_ShellPrmt18	0	0.090909091	0	0	0.006329114	0	0
Num_Facc19	**0**	**0.181818182**	**0.3333333**	**0**	**0**	**0**	**0**
Num_OBndCMD20	0	0	0	0	0	0	0
Is_HotLogIn21	**0**	**0**	**0**	**0**	**0**	**0**	**0**

(continued)

Table 6 (continued)

Feature name	apache2	buffer_overflow	ftp_write	guess_passwd	HTTPtunnel	imap	ipsweep
IsGust_LogIn22	0	0	0.3333333	0.022899015	0	0	0
Host_Count23	**43.10831234**	**1.227272727**	**1**	**1.033890543**	**1.240506329**	**1**	**6.003267974**
Srv_Count24	43.12846348	1.227272727	1	1.021983055	1.075949367	1	9.937908497
Syn_ErrR25	**0.317267003**	**0**	**0**	**0.000915961**	**0**	**0**	**0**
SrvSyn_ErrR26	0.317267003	0	0	0.000915961	0	0	0
Rej_ErrR27	**0.674017632**	**0**	**0**	**0.008090222**	**0.798481013**	**1**	**0**
SrvRej_ErrR28	0.674017632	0	0	0.007059766	0.860759494	1	0
Same_SrvR29	**1**	**1**	**1**	**0.994247767**	**0.930316456**	**1**	**1**
Diff_SrvR30	0	0	0	0.010648042	0.119303797	0	0
Srv_diffHostR31	**0.000151134**	**0**	**0**	**0.002976872**	**0**	**0**	**0.156862745**
Dst_HCount32	254.5579345	171.8636364	203.33333	244.557133	217.1708861	255	5.620915033
Dst_HsrvCount33	**250.418136**	**28.81818182**	**9.6666667**	**219.6910923**	**17.75949367**	**1**	**107.9248366**
SmDstH_SmSrvc34	0.983690176	0.319545455	0.0466667	0.891284635	0.086139241	0	0.990359477
DstH_DiffSrvR35	**0.005629723**	**0.153181818**	**0.0333333**	**0.0158049**	**0.027531646**	**0.02**	**0.000326797**
DstH_SmSrcPortR36	0	0.188636364	0.0166667	0.001880009	0.006392405	0	0.990359477
DstHSrvc_DiffHstR37	**0**	**0.046363636**	**0**	**0.002138768**	**0.000126582**	**0**	**0.894183007**
DstHst_SErrR38	0.12861461	0.072272727	0.3466667	0.000737348	6.33E-05	0	0
DstHstSrvc_SErrR39	**0.128916877**	**0.043636364**	**0**	**0.000904511**	**0**	**0**	**0**
DstHst_RSTerrR40	0.401725441	0.163181818	0.0133333	0.018234486	0.069177215	0	0
DstHstSrvc_RSTer41	**0.407002519**	**0.088636364**	**0**	**0.007563545**	**0.837658228**	**1**	**0**

Table 7 Mean, minimum and maximum values of each attack behavior in test set of KDD Cup'99

Value	Mean	Min	Max	Value	Mean	Min	Max
apache2	**962.987088**	**13.06119**	**2623.51024**	**pod**	**36.16149**	**15.7381**	**42.63571**
back	1493.087395	1055.85714	1510.5731	portsweep	7.75553	1.70381	10.7619
buffer_overflow	**510.9671636**	**24.82405**	**5379.05405**	**Processtable**	**134.6345**	**10.57595**	**206.141**
ftp_write	34.21722	3.59881	81.53714	ps	52.63381	8.5219	161.2845
guess_passwd	**16.46652526**	**4.74048**	**73.66881**	**rootkit**	**1143.010788**	**7.29071**	**6755.50619**
HTTPtunnel	25.56876	2.48929	2248.459	Saint	16.77856	1.81143	40.38143
imap	**48.42905**	**0**	**48.42905**	**Satan**	**17.55883**	**2.79619**	**20.78524**
ipsweep	4.591566699	1.28571	61.92857	sendmail	101.8643406	2.21429	313.70571
land	**2.845555556**	**1.16667**	**7.1881**	**smurf**	**54.7392626**	**26.07143**	**62.40476**
loadmodule	42.530715	0	42.85881	snmpgetattack	20.23294	8.365	123.0512
mailbomb	**94.68063199**	**2.02381**	**6765.4631**	**snmpguess**	**15.59099709**	**3.88095**	**33.30976**
Mscan	8.095948	1.58333	40.40571	Sqlattack	110.04	108.79167	111.28833
multihop	**135.3018**	**8.33405**	**1427.23**	**teardrop**	**8.781488333**	**3.4819**	**13.96571**
named	1352.392464	0.95238	14492.52381	udpstorm	3.92643	2.42857	5.42429
Neptune	**12.49666**	**3.1669**	**21.19048**	**Warezmaster**	**123.0512**	**2.02381**	**6770.31**
Nmap	4.422599	2.14881	6.83333	worm	293.21619	292.145	294.28738
normal	**115.1369**	**149804.9**	**1.80786**	**xlock**	**255417.562**	**15.71881**	**1498020**
perl	75.33429	72.58	78.08857	xsnoop	231.2451	63.12143	479.1198
phf	**107.7749**	**2.0519**	**213.4979**	**Xterm**	**392.2029669**	**19.53071**	**1075.79048**

This subsection presents a measure of central tendency for features and network's traffic record around their mean. Minimum, maximum and the mean values for features and traffic records have been presented here.

c. *Standard Deviation of Each Feature/Attribute*

Standard deviation (σ) is the measure of spread of data which is most commonly used in statistical practice. For feature's value, v_i mean of a feature is $\bar{\mu}$ of the total number of N items (number of rows) and the standard deviation (σ) will be as per Eq. (2). The standard deviation explores dispersion of data around the mean values of the features, and it never gets negative values.

$$\sigma = \sqrt{\frac{1}{N}\sum_{i=1}^{N}(v_i - \bar{\mu})^2} \tag{3}$$

A feature/attribute selected at random from a dataset whose σ Is low has a better chance of being close to the mean value than that of an item whose σ is higher.

Features 1, 5, 6, 23, 24, 32 and 33 as shown in Table 8 have high values that means that values of these feature are largely spread out which is the cause of outlier of the feature class in the dataset.

d. *Linear Correlation Coefficient (LCC)*

The fast, most popular and accurate method for describing the association between two random linear-dependent variables is correlation coefficient, and it is used for measuring the correlation among relationship between random-dependent variables, hence also called linear correlation coefficient [3].

If μ is the behavior of the traffic data record and v_i are other features in the traffic data record, then the correlation between these variables is according to Eq. 4.

$$\Gamma = \frac{\sum_{i=1}^{n}(\mu_i - \bar{\mu})(v_i - \bar{v})}{\sqrt{\sum_{i=1}^{n}(\mu_i - \bar{\mu})}\sqrt{\sum_{i=1}^{n}(v_i - \bar{v})}} \tag{4}$$

The linear correlation coefficient, Γ, can only take values between –1 and +1. A value of +1 indicates perfect positive relationship, a value of –1 indicates perfect negative relationship, and a value of 0 indicates no linear relationship.

Linear correlation coefficient, Γ, for 'Behaviour42' feature with all 42 features has been calculated which shows that there is a significant relation between traffic data behavior and those features whose Γ value is $\geq$ 50% as shown in Table 9. The negative sign of Γ reveals that there is a relation between traffic data behavior and the other features in opposite direction. If the value of $\Gamma 2$ is less than 50%, it means that there is no significant relationship between the traffic data behavior and the features of the dataset.

Correlation coefficient, Γ, for Behaviour42 feature with all 42 features has been shown in Table 9 which explores that the features 2, 3, 4, 12, 23, 24, 29, 33, 34 and 36 have a significant linear relationship with the behavior of the traffic data.

Table 8 Standard deviation of test set of KDD Cup'99 dataset

Feature's name	Std_Deviation	Feature's name	Std_Deviation
Duration1	407.36121	IsGust_LogIn22	0.049132302
Proto_Type2	0.64597064	Host_Count23	219.84192
Service3	14.587976	Srv_Count24	239.3129
Flag4	2.8248	Syn_ErrR25	0.23389307
Src_Byte5	127648.74	SrvSyn_ErrR26	0.234816
Dst_Byte6	16107.354	Rej_ErrR27	0.3468776
Land7	0.005379126	SrvRej_ErrR28	0.34772438
Wrng_Frgment8	0.040360767	Same_SrvR29	0.37121475
Urgnt9	0.009820883	Diff_SrvR30	0.10674749
Hot10	0.31184456	Srv_diffHostR31	0.12520674
Failed_LogIn11	0.049946561	Dst_HCount32	60.99437
Logged_In12	0.37818104	Dst_HsrvCount33	100.327
Num_Compro13	1.9583209	SmDstH_SmSrvc34	0.38723025
Root_Shell14	0.014115977	DstH_DiffSrvR35	0.096218541
su_Attempt15	0.005946902	DstH_SmSrcPortR36	0.4914397
Num_Root16	2.1651835	DstHSrvc_DiffHstR37	0.035866082
Num_FCreations17	0.19311617	DstHst_SErrR38	0.23125601
Num_ShellPrmt18	0.012929568	DstHstSrvc_SErrR39	0.23295595
Num_Facc19	0.029443378	DstHst_RSTerrR40	0.3437019
Num_OBndCMD20	0	DstHstSrvc_RSTerrR41	0.34610817
Is_HotLogIn21	0.006211183	Behviour42	6.6665864

e. *Analysis of Variance (ANOVA)*

Analysis of variance (ANOVA) is a statistical procedure for analyzing whether variation in the response variable arises within or among different population groups. Statistics and Machine Learning Toolbox in MATLAB provides one-way, two-way and N-way analysis of variance (ANOVA).

ANOVA is a kind of parametric method for means comparison and is an extension of t-test. It is a collection of statistical models which is used to analyze the differences among feature's means when there are more than two features to be compared and pair-wise t-test is not appropriate. ANOVA is used to summarize the strength of the association between a predictor(s) and the dependent variable which requires normality and equal variance. Otherwise, nonparametric analysis should be used.

ANOVA test has been used for analyzing the difference in the feature's means by partitioning the total variation in the data into two components

(a) Variation of feature means from the overall mean; that is, $\left(\overline{Y_j} - \overline{Y}_{all}\right)$ is the variation between features where $\overline{Y}_j$ is sample mean of feature j and $\overline{Y}_{all}$ is overall sample mean.

Table 9 Correlation coefficient for behavior feature with all 42 features

Feature name	Feature's label	Behviour42	Feature name	Feature's label	Behviour42
Duration1	1	−0.020114472	IsGust_LogIn22	22	0.035501726
Proto_Type2	2	−0.66996986	Host_Count23	23	0.66656834
Service3	3	−0.70691568	Srv_Count24	24	0.75349498
Flag4	4	0.53355277	Syn_ErrR25	25	−0.29065177
Src_Byte5	5	0.00554932	SrvSyn_ErrR26	26	−0.28923216
Dst_Byte6	6	−0.042590518	Rej_ErrR27	27	−0.45707425
Land7	7	−0.010980505	SrvRej_ErrR28	28	−0.45422059
Wrng_Frgment8	8	−0.007968064	Same_SrvR29	29	0.53526181
Urgnt9	9	0.007668075	Diff_SrvR30	30	−0.15715963
Hot10	10	−0.06005558	Srv_diffHostR31	31	−0.18744265
Failed_LogIn11	11	−0.12106624	Dst_HCount32	32	0.24374497
Logged_In12	12	−0.49899334	Dst_HsrvCount33	33	0.56251353
Num_Compro13	13	−0.007760417	SmDstH_SmSrvc34	34	0.56400263
Root_Shell14	14	−0.00959559	DstH_DiffSrvR35	35	−0.15140298
su_Attempt15	15	−0.004076816	DstH_SmSrcPortR36	36	0.85073179
Num_Root16	16	−0.002082285	DstHSrvc_DiffHstR37	37	−0.11481885
Num_FCreations17	17	−0.004232858	DstHst_SErrR38	38	−0.28617293
Num_ShellPrmt18	18	−0.001824248	DstHstSrvc_SErrR39	39	−0.28442317
Num_Facc19	19	−0.019855412	DstHst_RSTerrR40	40	−0.4519603
Num_OBndCMD20	20	0	DstHstSrvc_RSTer41	41	−0.45014802
Is_HotLogIn21	21	0.002379089	Behviour42	42	1

(b) Variations of observations in each feature's value from their feature's mean estimate variation within feature column; that is, $\left(Y_{ij} - \overline{Y}_j\right)$ where $\overline{Y}_{ij}$ is ith element or feature's value in jth feature and $\overline{Y}_j$ is sample mean of feature j.

ANOVA partitions the total sum of squares (SST) into sum of squares due to between-feature's effect (SSR) and sum of square error (SSE) (The MathWorks 1994-2018).

$$SST = \sum_i \sum_j \left(Y_{ij} - \overline{Y}_{all}\right)^2 = \sum_j n_j \left(\overline{Y_j} - \overline{Y}_{all}\right)^2 + \sum_i \sum_j \left(Y_{ij} - \overline{Y}_j\right)^2 \quad (5)$$

where

$$SSR = \sum_j n_j \left(\overline{Y_j} - \overline{Y}_{all}\right)^2 \quad (6)$$

$$SSE = \sum_i \sum_j \left(Y_{ij} - \overline{Y}_j\right)^2 \quad (7)$$

n_j is the sample for the jth feature, j = 1, 2, 3,….k and k = 41.

ANOVA compares the variation 'between-features' to the variation 'within-features.' If the ratio of within-group variation to between-group variation is significantly high, then the group means are significantly different from each other. Measurement of this can be performed using a test statistic called F-distribution or F-statistic, which is the ratio of the mean squares with (N – 1, /k – N), degrees of freedom.

$$F = \frac{SSR/(K-1)}{SSE/(N-k)} \tag{8}$$

where N is total numbers of observations, i.e. (numbers of rows* numbers of columns), k is the numbers of features in network traffic data, (N − k) is degree of freedom (df) for within-features (error), (k − 1) is df for 'between-features' (columns) and (N − 1) is total degree of freedom which is equal to (N − k) + (k − 1).

Two-way ANOVA is used to comparing the means of two or more columns and two or more rows of the test set of KDD Cup'99. In this module, we see the significance of the features/columns in the offline KDD Cup'99 dataset on behavior and other features of traffic data.

ANOVA Table

The ANOVA table is used to define the variability in the model by source. The F-statistic has been used for testing the significance of the variability, while the p-value is decided on the significance of this variability.

The primary purpose of a two-way ANOVA is to understand the interaction between the rows or columns of the test set. In ANOVA test set, it has been observed that sum of squares (SS) due to each source, degree of freedom (df) for columns, mean squares (MS = SS/df) for each source, F-statistic and p-value are given in Table 10. Here, p-value is the probability that the F-statistic can take a value larger than the computed test statistic value. Errors in Table 10 define the variability due to the differences between the data in each feature/attribute and the feature/attribute mean. This variability is also called as variability within groups. The total number of columns as features, say k, are 41 with a behavior column which is not taken in the ANOVA. Because the behavior column as 42nd column just tells about the behavior of traffic data. Total number of elements, N, in the test set are 311029*41 = 12752189. The degree of freedom (df) for columns/features is (41-1) equal to 40.

Table 10 Two-way ANOVA table of the traffic data in corrected (test set) of KDD Cup'99 dataset

Source	SS	Df	MS	*F*	Prob > *F*
Columns	1.08475e + 12	40	2.71187e + 10	67.16	0
Rows	1.26204e + 14	311028	4.05764e + 08	1	0.0271
Error	5.0233e + 15	12441120	4.03766e + 08		
Total	5.15058e + 15	12752188			

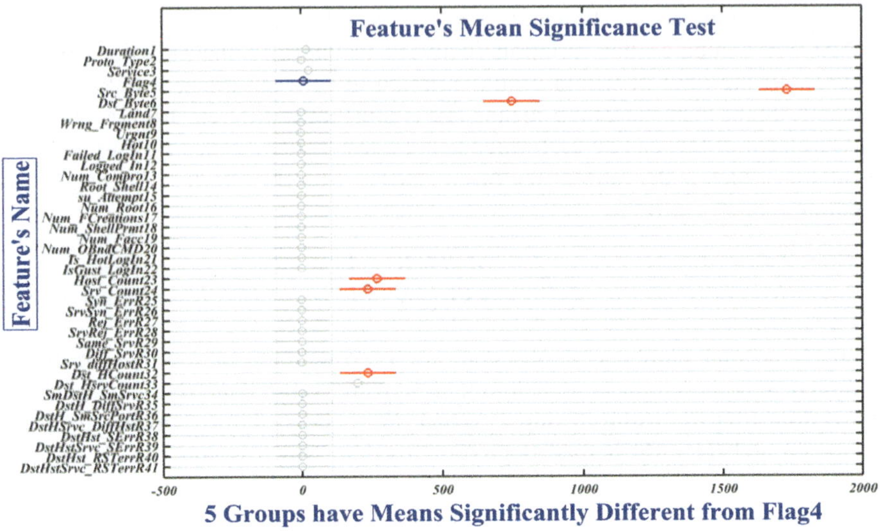

Fig. 3 Fourth feature mean significance test

F-statistic as shown in Table 10 has a value equal to 1, which means there is no statistical significance of row-wise mean in the dataset while column-wise mean has statistical significance with a large F-statistic value, 67.16, in Table 10. The p-value represents the probability that the outcomes are of a chance or have statistical significance. Value $p < 0.05$ indicates that result is 'statistically significant' and not a matter of chance alone. According to Winters et al. in [26], if the outcome of p is less than 5%, observed result is not by chance in traffic dataset.

f. *Feature's Mean Significance Test*

Feature's mean significance test has been shown in Figs. 3, 4, 5, 6 and 7. These figures describes that the group of features in red colour are 'means significantly different' from the feature in blue color. While, features in gray color do not have mean significance difference with feature in red color.

Figure 3 shows that a group of five features labeled as 5, 6, 23, 24 and 32 according to Table 9 have means significantly different from feature having label as 4, and feature's name is Flag4. Similarly, the same group of five features, labeled as 5, 6, 23, 24 and 32, have mean significant difference from the features labeled as 1, 2, 3, 29, 34 and 36.

Figure 4 shows that a group of forty features labeled as 1–4, 6–41 according to Table 9 have means significantly different from feature having label as 5, and feature's name is Src_Byte5.

Figure 5 shows that a group of forty features labeled as 1–5, 7–41 according to Table 9 have means significantly different from feature having label as 6, and feature's name is Dst_Byte6. Similarly, a group of six features labeled as 5, 6, 23, 24, 32 and 33 according to Table 9 have means significantly different from feature having label

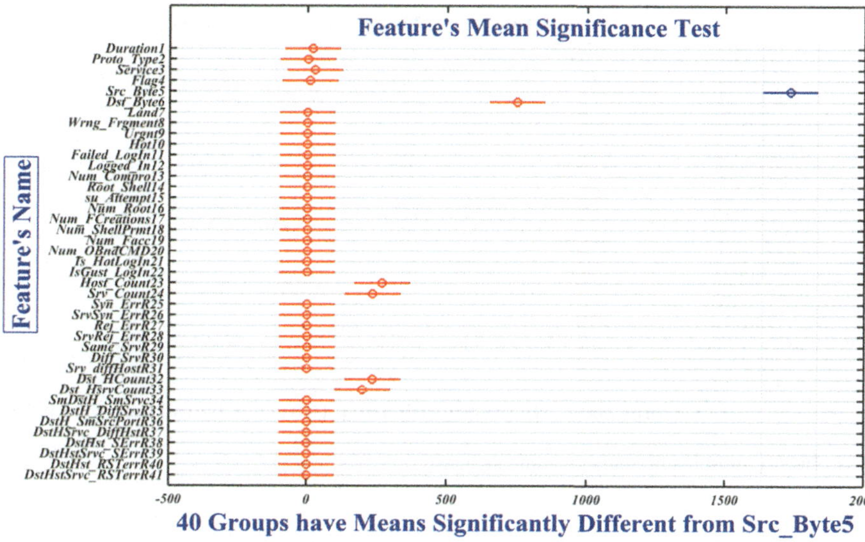

Fig. 4 Fifth feature mean significance test

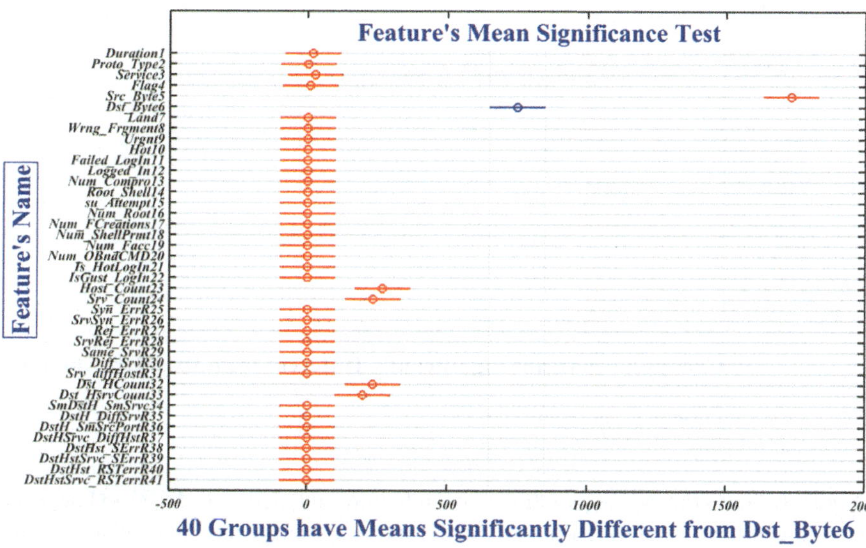

Fig. 5 Sixth feature mean significance test

as 18, and feature's name is Num_ShellPrmt18. Moreover, the same group of six features, labeled as 5, 6, 23, 24, 32 and 33, have mean significant difference from the features labeled as 7–22, 25–28, 30, 31, 35, 37–40.

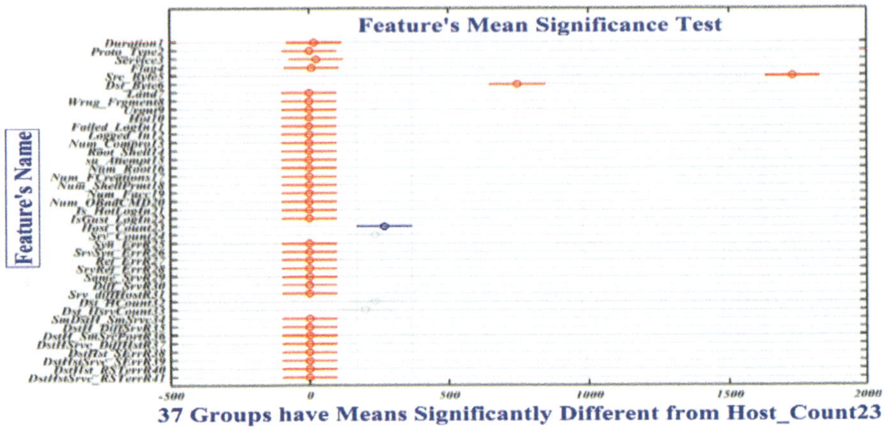

Fig. 6 Twenty-third feature mean significance test

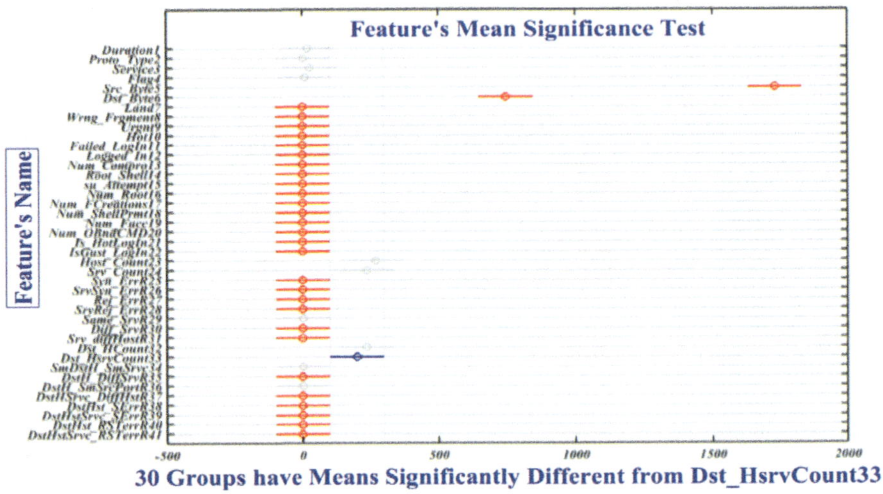

Fig. 7 Thirty-third feature mean significance test

Figure 6 shows that a group of thirty-seven features labeled as 1–22, 25–31 according to Table 9 have means significantly different from feature having label as 23, and feature's name is Host_Count23. Similarly, the same group of thirty-seven features, labeled as 1–22, 25–31, have mean significant difference from the features labeled as 24 and 32 also.

Figure 7 shows that a group of thirty features labeled as 5–22, 25–28, 30, 31, 35, 37–41 as described in Table 9 have means significantly different from feature having label as 33, named as Dst_HsrvCount33.

Table 11 Mean significance test table

S. no.	Features with label as in Table 9	Numbers of features in group having mean significant difference	Feature which are included in a group
1.	1, 2, 3, 4, 29, 34, 36	05	5, 6, 23, 24, 32
2.	7 to 22, 25 to 28, 30, 31, 35, 37 to 41	06	5, 6, 23, 24, 32, 33
3.	33	30	5 to 22, 25 to 28, 30, 31, 35, 37 to 41
4.	23, 24, 32	37	1 to 22, 25 to 31, 34 to 41
5.	5	40	1 to 4, 6 to 41
6.	6	40	1 to 5, 7 to 41

Table 11 shows that subset of features that are shown in the fourth column of the table is significantly different from the features that are given in the second column of this table. Instead of feature's name, feature's labels are given in Table 11 as per Table 9.

7 Conclusion

This paper analyzes that how the KDD Cup'99 dataset is distributed and organized. This statistical analysis outlines the structure of the KDD Cup'99 dataset used in this work. The various types of features/attributes of the traffic data and their aspects are studied with a focus on their role in the behavior of network traffic data and in intrusion detection system. The basic components, structures and organization of network traffic records with their feature types are shown in Table 1. Attack categorization is shown in Table 2. Feature's label with behavioral record statistics of KDD Cup'99 has also been explicitly explained in Table 4. This work also explains the mean values of every feature along with each type of behavior of the traffic records. Mean value for the same feature of different behaviors also shows that how the data of same features is spread out with central tendency. It also provides the visualization of feature's mean value for training and test sets of KDD Cup'99 dataset. The correlation between features of dataset has been given in Table 9. Hence, the association between the features and the traffic record behavior has been measured in Table 9. Table 10 represents the interaction between the rows or columns of the test set. It shows that there is no statistical significance of row-wise mean in the dataset while column-wise mean has statistical significance in the dataset. The F-statistic in ANOVA table shows the significance of the variability. Table 11 presents mean significance test that generates the subset of features which is significantly different from other features of the dataset.

References

1. https://in.mathworks.com/help/stats/two-way-anova.html
2. Ali Z, Bhaskar SB (2016) Basic statistical tools in research and data analysis. Indian J Anaesthesia 60(9):662
3. Ambusaidi MA, He X, Nanda P, Tan Z (2016) Building an intrusion detection system using a filter-based feature selection algorithm. IEEE Trans Comput 65(10):2986–2998
4. Anderson JP (1980) Computer security threat monitoring and surveillance. In: Technical Report, James P. Anderson Company, pp 1–56
5. Denning DE (1987) An intrusion-detection model. IEEE Trans Softw Eng 2:222–232
6. Horng S-J, Su M-Y, Chen Y-H, Kao T-W, Chen RJ, Lai J-L, Perkasa CD (2011) A novel intrusion detection system based on hierarchical clustering and support vector machines. Expert systems Appl 38(1):306–313
7. Stolfo J, Fan W, Lee W, Prodromidis A, Chan PK (2000) Cost-based Modeling and Evaluation for Data Mining With Application to Fraud and Intrusion Detection. http://kdd.ics.uci.edu/databases/kddcup99/task.htmlAccessed 22 May 2018
8. Kalavadekar MPN, Sane SS (2017) Effective intrusion detection systems using genetic algorithm. IJETT 1(2):8316–8319
9. Karthikeyan KR, Indra A (2010) Intrusion detection tools and techniques-a survey. Int J Comput Theory Eng 2(6):901–906
10. Kavousi F, Akbari B (2012) Automatic learning of attack behavior patterns using Bayesian networks. In: Sixth international symposium in telecommunications (IST), Tehran, Iran: IEEE, pp 999–1004
11. Kayacik HG, Zincir-Heywood AN, Heywood MI (2005) Selecting features for intrusion detection: a feature relevance analysis on KDD 99 intrusion detection datasets. In: Proceedings of the third annual conference on privacy, security and trust
12. Kayacik HG, Zincir-Heywood AN, Heywood MI (2003) On the capability of an SOM based intrusion detection system. In: Proceedings of the International Joint Conference on Neural Networks, vol 3. Portland, OR, USA, IEEE, pp 1808–1813
13. Kendall KKR (1999) A database of computer attacks for the evaluation of intrusion detection systems. Doctoral dissertation, Massachusetts Institute of Technology
14. Lakshmi TVN, Babu VK (2015) Detection of user to root attacks using machine learning techniques. Int J Adv Eng Global Technol 3(3):418–424
15. Lee W, Stolfo SJ (2000) A framework for constructing features and models for intrusion detection systems. ACM Trans Inf Syst Secur 3(4):227–261
16. Lippmann R, Haines JW, Fried DJ, Korba J, Das K (2000) The 1999 DARPA off-line intrusion detection evaluation. Comput Netw 34(4):579–595
17. Tavallaee M, Ebrahim E, Lu W, Ghorbani AA (2009) a detailed analysis of the KDD CUP 99 data set. In: IEEE symposium on Computational Intelligence in Security and Defense Applications, (CISDA 2009), Ottawa, ON, Canada: IEEE, pp 1–6
18. McHugh J (2000) Testing intrusion detection systems: a critique of the 1998 and 1999 darpa intrusion detection system evaluations as performed by lincoln laboratory. ACM Trans Inf Syst Secur (TISSEC) 3(4):262–294
19. Moustafa N, Slay J (2016) The evaluation of network anomaly detection systems: statistical analysis of the UNSW-NB15 data set and the comparison with the KDD99 data set. Inf Secur J: A Global Perspective 25(1–3):18–31
20. Moustafa N, Creech G, Slay J Big data analytics for intrusion detection system: statistical decision-making using finite dirichlet mixture models. In: Data analytics and decision support for cybersecurity, Springer, Cham (Springer International Publishing), pp 127–156
21. Mukkamala S, Janoski G, Sung A (2002) Intrusion detection using neural networks and support vector machines. In: Proceedings of the 2002 International Joint Conference on Neural Networks, IJCNN'02, vol 2. Honolulu, HI, USA, IEEE, pp 1702–1707
22. Özgür A, Erdem H (2016) A review of KDD99 dataset usage in intrusion detection and machine learning between 2010 and 2015. PeerJ PrePrints 4:1–21

23. Patel SK, Mehta UH, Patel UM, Bhagat DH, Nayak P, Patel AD (2015) A technical review on intrusion detection system. Int J Comput Sci Eng Technol 6(01):17–25
24. Sharma C, Jain SC, Sharma AK (2016) Explorative study of SQL injection attacks and mechanisms to secure web application database-A. Int J Adv Comput Sci Appl 7(3):79–87
25. Win MTM, Khaing KT (2013) Analyzing knowledge based feature selection to detect remote to local attacks. Int J Adv Res Comput Eng Technol 2(5):1762–1765
26. Winters R, Winters A, Amedee RG (2010) Statistics: a brief overview. Ochsner J 10(3):213–216

Efficient Certificateless Privacy-Preserving Public Auditing for Cloud-Based IIoT

Tapaswin Padhy and Syam Kumar Pasupuleti

Abstract Continuous development of the Industrial Internet of Things (IIoT) is making organization to adopt cloud for data storage and computation as it is cost-effective and less computation for IoT devices. Since cloud is untrusted, authentication of data stored in the cloud is one of the major concerns. To address this issue, several schemes are proposed using certificateless signatures but these schemes are inefficient. In this paper, we propose an efficient certificateless privacy-preserving public auditing for cloud-based IIoTs using certificateless short signature. Unlike previous schemes, our scheme has low computational cost for signature generation and verification. Further, our scheme enables the public auditor to verify the authentication of data and also protects the privacy of data from public auditor during verification. Security of this scheme is proved against Type-I, Type-II and Type-III adversaries. The performance analysis shows that our scheme is efficient and feasible for the IIoTs.

Keywords Industrial Internet of Things (IIoT) · Certificateless signature (CLS) · Authentication · Privacy-preserving cloud computing · Bilinear pairing

1 Introduction

Internet of Things (IoT) technology changes the day-to-day life of people in the future. The IoT devices [1] can connect each other through a wireless network technology for data transfer. Since IoT devices are widely used in the industrial field, the industrial IoT (IIoT) [2] appears. The IIoT can greatly improve manufacturing efficiency, reduce product costs and resource consumption [3]. By considering the huge amount of data generated by IIoT, the storage and processing of such data are a great

T. Padhy (✉)
International Institute of Information Technology, Bhubaneswar, India
e-mail: taps1197@gmail.com

S. K. Pasupuleti
Institute for Development and Research in Banking Technology, Hyderabad, India
e-mail: psyamkumar@idrbt.ac.in

S. M. Thampi et al. (eds.), *Applied Soft Computing and Communication Networks*, Lecture Notes in Networks and Systems 125,
https://doi.org/10.1007/978-981-15-3852-0_10

challenge. Cloud storage [4] is the best option to store huge data, because the cloud has enough storage and powerful processing capacity to store and processing of data.

The converge of IIoT and cloud brings many benefits to users, but how to guaranty the integrity of data in the cloud is one of the major concerns for adoption of the cloud-assisted IIoT [5]. Because, cloud is untrusted and malicious adversary may potentially modify or inject malicious data can seriously effect both the data owners and data users. Hence, efficient verification mechanism is required to verify the of authentication of data in the cloud. In this mechanism, data should be preprocessed before sending to cloud and integrity will be verified later.

Organization: "Motivation in given Sect. 1.1, related work in Sect. 1.2, and contributions are presented in Sect. 1.3. System model is in Sect. 2. Construction of proposed ECLPPPA is given in Sect. 3. Security analysis is presented in Sect. 4, and performance analysis given in Sect. 5. Conclusion is given in Sect. 6".

1.1 Motivation

Due to limited storage and computing power, data collected by IIoT is outsourced to cloud. However, authentication outsourced data is major concern due to the following reasons: (1) The user lose physical control over data once data moves to the cloud, so users cannot verify the authentication of their data. (2) Although cloud provides powerful and reliable infrastructures for data storage than local storage devices, they are more vulnerable to both internal and external attacks due to shared technology. (3) Sometimes, malicious CSPs may try to hide data corruption or data loss incidents for their own benifits. Hence, users require efficient and secure authentication verification schemes to get the assurance of their data in cloud.

1.2 Related Works

To verity authentication of outsourced data, "public key infrastructure (PKI)"-based authentication scheme proposed [6], in this scheme, the certificate management creating huge overhead that is given by a certificate authority. To address this issue, identity (ID)-based authentication schemes are proposed [7] where private key of user is created based on his own identity that eliminates the necessity of the certificate management. However, ID-based schemes suffer from key escrow problem, i.e., KGC can forge the signature of any user easily because it generates user private key. To solve this problem, certificateless-based authentication scheme (CLS) was proposed by Al-Riyami and Paterson's [8]. In this scheme, the private key of the user consists of two parts: the partial private key, which is generated by KGC and the user secret key, which is generated by the user itself. Since, the KGC does not have complete knowledge of the user secret key, the key escrow problem in the ID-based signature can be easily eliminated. Then, the certificate management is also unnecessary in the CLS, because the public key is also generated by user itself. Based on this scheme [8], a number of certificateless signatures schemes were developed [9–24] to verity the message authentication.

The CLS schemes [9, 10] proposed based on bilinear pairings to reduce the number of pairing operations in sign and verify algorithms. Later, Zhang et al. [11] designed a new "certificateless signature (CLS)" using bilinear pairing. They achieved tight security reduction under CDH assumption in ROM but it is not efficient. Huang et al. [12] devised a "new CLS scheme, which produces enough short-size signature." Shim [13] exhibited that [12] plan is unreliable against Type-I adversary. From that point, Tso et al. [14] suggested another CLSS scheme. Despite the fact that they have demonstrated that their plan is provably safe under k–CAA presumption, but it appeared unreliable against Type-I adversary. Choi et al. [15] proposed another CLS plan utilizing bilinear pairing. Furthermore, Yu et al. [16] proposed an enhanced CLS scheme that has many benefits like "shorter system parameters and higher computational efficiency" than the previous CLS schemes. Unfortunately, Yuan and Wang [17] showed that the scheme in [16] is insecure against Type-2 adversary. Besides, they also designed a new CLS scheme, and it was constructed based on bilinear pairing. As we know, all schemes [9–17] are proposed based on bilinear pairing, which are thought to be the time expensive cryptographic activity. To improve the better efficiency, He et al. [18] presented a new CLS scheme without pairings. Although it was a good idea to make their scheme more efficient for low computing resources, next Tian et al. [19] showed that the CLS scheme in [18] is insecure against Type-2 adversary. Later, many improved CLS schemes proposed without pairings [20, 21]. However, in pairing-free CLS schemes, random numbers are needed for signature generation, so these schemes always have a larger size and are not suitable for IIoT devices due to low bandwidth and limited storage capacity.

Recently, Karati et al. [22] presented a lightweight CLS scheme for IIoT environments without map to point hash functions. This scheme is secure against both the Type-I and Type-II adversaries and reduced the computational cost compared to other CLS schemes. However, this scheme failed to provide the appropriate security. Hu et al. [23] proposed a "pairing-free and provably secure certificateless parallel key-insulated signature" scheme for IIoT environments that reduces secret key disclosure risk in the IIoT environment. However, this creates computation overhead for data consumer to verify the authenticity of data. He et al. [24] proposed a certficateless public auditing scheme for cloud-based wireless body area networks (WBAN). In this scheme, the public verifier verifies the authenticity of message to reduce the burden of data consumer. Although this scheme secure against both Type-1 and 2 adversaries, it did not protect privacy of data from public verifier (Type-3), i.e., the verifier may access information about data content during verification.

1.3 Contributions

In this paper, we propose an efficient certificateless privacy-preserving public auditing (ECLPPPA) scheme for cloud-based IIoT using certificateless short signatures. The main contributions of ECLPPPA are as follows:

1. ECLPPPA enables the public auditor to verify the integrity of data in cloud and preserves the privacy of data from auditor during verification process.
2. ECLPPPA scheme reduces the computation overhead of signer and verifier, because it requires one multiplication point operation and one pairing operation during signing and verification process.
3. ECLPPA is secure against the Type-I, II and III adversaries in random oracle model.

2 System Model

2.1 System Architecture

The system architecture of ECLPPPA as shown in Fig. 1 consists of four entities:

Users: The users collect data, which is generated by IIoT devices and stores in cloud. Later, request auditor for integrity verification of data stored in cloud.

KGC: Its responsibility is to generate public parameters and also to generate the user's partial key based on the identity of user.

Third-Party Auditor (TPA): It is an inquisitive third party, and it is supposed to check data authentication in cloud server. As user sends an audit request to TPA, upon receiving audit request, TPA checks data integrity and return status to user.

Cloud Server: It is a semi-trusted third party with scalable storage and computation power to handle the outsourced data.

2.2 Adversary Model

Here, we consider three types of adversaries: Type-I adversary A_1, which cannot get to the secret key, but can displace the users's public key, Type-II adversary A2, which cannot override any users's public key, but can get to the user secret key and Type-III adversary A3, which can get access to data while data auditing is happening.

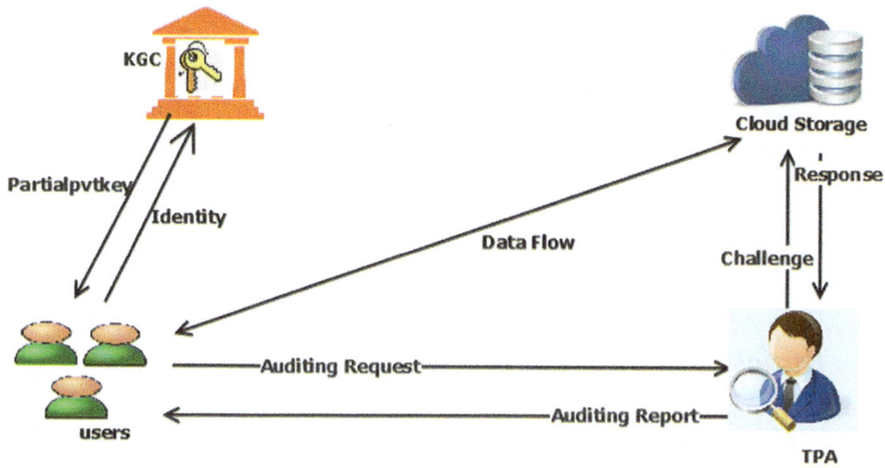

Fig. 1 System architecture

2.3 Design Goals

The main design goals for our scheme are as follows:

1. **Public Auditing**: TPA checks the correctness for data in the cloud without getting the whole data or certainly any extra burden on data users.
2. **Privacy Preserving**: TPA cannot access the data while auditing process.
3. **Efficient**: Computation cost for signature generation and verification should be minimum.

3 Construction of ECLPPPA

The construction of ECLPPPA consists of seven algorithms: Setup, Partial Private Key Extract, Public Key Gen, Private Key Gen, Tag-Gen, Proof-Gen and Proof-Verify. We describe each of them below.

3.1 Setup

This algorithm generates public parameters and master secret key as follows:

1. For given security parameter, KGC chooses two groups: Let G_1 be an additive, whereas G_2 be a multiplicative group of similar order q and $e : G_1 \times G_1 \rightarrow G_2$ be a bilinear pair.

2. Say $D \in G_1$ is the generator for group G_1.
3. Choose a large prime number q.
4. Let H_1 be cryptographically secure hash functions such that $H_1 : \{0, 1\}^* \times G_1 \times G_1 \rightarrow Z_q^*$.
5. Let $h : \{0, 1\}^* \rightarrow Z_q^*$ be a secure hash function.
6. KCG randomly chooses $s \in Z_q^*$ as its master secret key and computes $P_{\text{pub}} = s \cdot D$ as it's public key.
7. KCG then publishes the parameters $\{q, G_1, G_2, D, h, H_1, e, s, P_{\text{pub}}\}$.

3.2 Partial Private Key Extract

In this algorithm, KGC generates partial private key based on public parameters, master secret key and a user identity ID as follows:

1. User sends unique identity ID_u to KGC.
2. KGC computes partial private key $S_u = s \cdot H_u$ where $H_u = h(\text{ID}_u)$ and return it to the user.
3. KGC send partial private key S_u to the user.

3.3 Set Secret Value

Given public parameters, the user chooses a random number x_{ID} with identity ID as secret value

1. User sets $sk_u = S_u, x_u$ as its private key.

3.4 Public Key Gen

This algorithm generates public key for user based on public parameters, user identity and secret value as follows:

1. User randomly selects $x_u \in Z_q^*$.
2. Then user computes $P_u = x_u \cdot D$ and sets P_u as it's public key.

3.5 SignGen

This algorithm generates signatures for data blocks based on user private key.

1. User divides file $M = \{m_1, m_2, \ldots, m_n\}$ is into n blocks.

2. For each user O with identity ID_o, a random number $r_o \in G_1$.
3. User calculates.

$$Q = H_1(\mathrm{ID}_O, P_{\mathrm{Pub}}, r_O) \cdot D$$

$$S_i = (s_o + x_o)(\mathrm{id}_i + m_i \cdot Q)$$

4. Returns $\{m_i, \mathrm{id}_i, S_i, r_o\}$ to Cloud, id_i is unique identity for each data element m_i.

3.6 *Challenge*

TPA creates a challenge to check authentication of data, and auditor chooses a random subset $J \subseteq [1, n]$ and a random number $a \in Z_q$ for challenge generation. Challenge $=$ $[\mathrm{ID}_o, a, J]$ and sends to cloud server.

3.7 *Proof-Gen*

Upon receiving challenge, the cloud server generates a proof.

1. Cloud generates a set $v = \{(i_j, w_j)\}_{j \in J}$ for every element $i_j \in J$. Here $w_j = a^j \bmod q$. Also, cloud picks a random number $b \in Z_q^*$.
2. The cloud server computes.

$$\sigma = \sum_{j \in J} w_j Sj$$

$$\mu = \sum_{j \in J} (w_j m_j + b \cdot H_1(\mathrm{ID}_j, B, P_{\mathrm{Pub}})) \quad \mathrm{mod} \ \cdot q$$

3. Finally, cloud returns $\{\sigma, \mu$ and $B\}$ to TPA as a proof, where $B = b \cdot D$.

3.8 *Proof-Verify*

Upon receiving $\{\sigma, \mu$ and $B\}$ from cloud, TPA verifies the proof using equation below

$$e(\sigma, D) = e\left(\sum_{j \in J} w_j \cdot \left(\mathrm{id}_j + Q \cdot \mu - Q \cdot B \cdot H_1(\mathrm{ID}_o, B, P_{\mathrm{pub}}) \right), P_{\mathrm{pub}} \cdot h(\mathrm{ID}_o) + P_o \right)$$

4 Security Analysis

This section shows that ECLPPPA is secured against Type-I, Type-II and Type-III adversaries.

Theorem 1 (Correctness) *The correctness of proposed scheme is proved by using equation as follows:*

$$e(\sigma, D) = e\Big(\sum_{j\in J} w_j \cdot \big(id_j + Q \cdot \mu - Q \cdot B \cdot H_1(ID_o, B, P_{\text{pub}})\big), P_{\text{pub}} \cdot h(ID_o) + P_o\Big)$$

$$\begin{aligned}
LHS &= e(\sigma, D)\\
&= e\Big(\sum_{j\in J} w_j \cdot S_j, D\Big)\\
&= e\Big(\sum_{j\in J} w_j \cdot \big((S_o + x_o) \cdot (id_j + m_j \cdot Q)\big), D\Big)\\
&= e\Big(\sum_{j\in J} w_j \cdot (id_j + m_j \cdot Q), (S_o + x_o) \cdot D\Big)\\
&= e\Big(\sum_{j\in J} w_j \cdot id_j + \sum_{j\in J} w_j \cdot m_j \cdot Q, (S_o \cdot D + x_o \cdot D)\Big)\\
&= e\Big(\sum_{j\in J} w_j \cdot id_j + \sum_{j\in J} w_j \cdot m_j \cdot Q, (s \cdot h_o \cdot D + x_o \cdot D)\Big)\\
&= e\Big(\sum_{j\in J} w_j \cdot id_j + \sum_{j\in J} w_j m_j Q + QbH_1(ID_o, B, P_{\text{pub}}) - QbH_1(ID_o, B, P_{\text{pub}})(P_{\text{pub}} \cdot h(ID) + P_o)\Big)\\
&= e\Big(\sum_{j\in J} w_j \cdot \big(id_j + Q \cdot \mu - Q \cdot B \cdot H_1(ID_o, B, P_{\text{pub}})\big), P_{\text{pub}} \cdot h(ID_o) + P_o\Big)\\
&= RHS
\end{aligned}$$

Theorem 2 *The proposed ECLPPPA is secure against the Type-I adversary if the CDH problem of the group G1 is hard.*

Proof Assume that there is a adversary A_1 that could break our scheme. Then, we can build a challenger S to solve the CDH problem with a non-negligible probability. Given an occasion $(D, Q_1 = a \cdot D, Q_2 = b \cdot D)$ of the CDH problem, A_1 sets $P_{\text{Pub}} \leftarrow Q_1$ and returns system parameters params $=\{q, G_1, G_2, D, e, H, P_{\text{Pub}}\}$ to A_1. S creates an identity ID_O as a challenge identity for A_1's questions as follows:

h query: S chooses a list L_h of triplet $\{\text{ID}_U, T_U, h_u\}$ starting with an empty list. When a request is made with message $\{\text{ID}_u, T_u\}$ S checks whether triplet is present in L_h. If it is found then S returns h_u to A_1; otherwise, S generates a random number $L_h \in Z_q$ stores in L_h to A_1.

H_1 query: S chooses a list L_{H_1} of triplet $\{\text{ID}_U, pk_u, P_{\text{Pub}}, Q\}$ starting with an empty list. When a request is made with message $\{\text{ID}_u, T_u\}$ S checks whether triplet is present

in L_{H_1}. If it is found then S returns Q to A_1; otherwise, S generates a random number $L_{H_1} \in Z_q$ stores in L_{H_1} to A_1.

Create User: S maintains an empty list L_k with initialized pairs $\{\mathrm{ID}_u, s_u, T_u, x_u, P_u\}$. after getting a query with identity ID_O, it checks type of pair is present in L_k, and its present returns pk_u to A_1. A invokes query for a identity ID, and the following steps are performed.

1. If $\mathrm{ID}_u = \mathrm{ID}_O$, S chooses two random numbers $t_u, x_u \in Z_q$ and set $s_u \leftarrow \perp$.
2. $T_u = t_u.D$ and $P_u = x_u.D$ and stores in the list L_k;
3. If $(\mathrm{ID}_u \neq \mathrm{ID}_O)$, S generates three random numbers $s_u, w_u, x_u \in Z_q$ and value of $T_u = s_u.D - w_u.P_{\mathrm{Pub}}$ and list L_k and L_h are updated as $\{\mathrm{ID}_U, s_U, T_U, x_U, P_U\}$ and $\{m, w_u\}$, respectively, $(m = \{\mathrm{ID}_U, T_U\})$.
4. S returns $pk_u = \{T_u, P_u\}$ to A_1.

Partial Private Key: On obtaining identity ID_U, S checks ID_U and ID_O equality. If equality is found, S aborts the game, else S checks L_k and returns $\{s_u, T_u\}$ to A_1.

Public Key Replacement: When query is received for user identity ID_U and $pk_u^{'} = \{T_U^{'}, D_U^{'}\}$, S looks up L_k and replaces $pk_u = \{T_U, P_U\}$ with $pk_u^{'}$. *Secret Value*: When query is received for user identity ID_U, S looks up L_K and returns x_u to A_1. *Tag-Gen*: When query is received for user identity ID_U and message m_i with its identity id_i, S raises query H with id_i and it returns a tuple.

1. If $c_i = 0$, S computes Tag S_i and returns $\{m_i, \mathrm{id}_i, S_i\}$ to A_1.
2. Else if $c_i = 1$, S aborts the game.

Finally, A_1 returns a signature $\sigma^{'}$ for ID on a message $m^{'}$. If signature can pass the proof verification equation, then we show the probability that A_1 successfully forges a signature as follows:

- M_1: S does not terminate set partial private key algorithms.
- M_2: A_1 successfully forges a signature σ on m for ID.
- M_3: The forged sign σ allows $\mathrm{ID}_O = \mathrm{ID}_O^{'}$.

From the test runs, we know

$$Pr[M_1] \geq \left(1 - \frac{q_{H_1}}{q}\right)^{q_{H_1}}$$

$$Pr[M_2|M_1] \geq \epsilon$$

$$Pr[M_3|M_1 \wedge M_2] \geq \frac{q_s}{q}$$

where A_1 sends maximum hash queries and sign queries. Then, the probability that S breaks the CDH problem is only:

$$Pr[M_1 \wedge M_2 \wedge M_3] = Pr[M_1].Pr[M_2|M_1].Pr[M_3|M_1 \wedge M_2] \geq \left(1 - \frac{q_{H_1}}{q}\right)^{q_{H_1}} \epsilon \frac{q_s}{q}.$$

Since ϵ is non-negligible, A_1 cannot break the CDH problem. Hence, the proposed ECLPPPA scheme is secure against Type-I adversary.

Theorem 3 *The proposed ECLPPPA scheme is secure against Type-II adversary in random oracle if Inv-CDH problem in G_1 is hard.*

Proof Assume there is a adversary A_2 that can break Inv-CDH problem in the game made by challenger S. Initially, the algorithm contains an empty hash list and an empty public hash list L_{H_1}, L_p, respectively.

Setup: In the following, algorithm A_2 takes a safe parameter k and returns the system **params** and KGC's master private key s to A_2.

h query: S chooses a list L_h of triplet $\{\text{ID}_U, T_U, h_u\}$ starting with an empty list. When a request is made with message $\{\text{ID}_u, T_u\}$, S checks whether triplet is present in L_h. If it is found, then S returns h_u to A_1; otherwise, S generates a random number $L_h \in Z_q$ stores in L_h to A_2.

H_1 query: S chooses a list L_{H_1} of triplet $\{\text{ID}_U, pk_u, P_{\text{Pub}}, Q\}$ starting with an empty list. When a request is made with message $\{\text{ID}_u, T_u\}$, S checks whether triplet is present in L_{H_1}. If it is found, then S returns Q to A_2; otherwise, S generates a random number $L_{H_1} \in Z_q$ stores in L_{H_1} to A_2.

Create User: S maintains a empty list L_k with initialized pairs $\{\text{ID}_u, s_u, T_u, x_u, P_u\}$. Upon receiving a query with identity ID_O, it checks type of pair is present in L_k, if it is present, it returns pk_u to A_2. A invokes query for a identity ID, and the following steps are performed.

1. If $\text{ID}_u = \text{ID}_O$ S chooses 2 random numbers $t_u, x_u \in Z_q$ and set $s_u \leftarrow \perp$.
2. $T_u = t_u.D$ and $P_u = x_u.D$ and stores in the list L_k;
3. If $(\text{ID}_u \neq \text{ID}_O)$ S generates three random numbers $s_u, w_u, x_u \in Z_q$ and value of $T_u = s_u.D - w_u.P_{\text{Pub}}$ and list L_k and L_h are updated as $\{\text{ID}_U, s_U, T_U, x_U, P_U\}$ and $\{m, w_u\}$, respectively, $(m = \{\text{ID}_U, T_U\})$.
4. S returns $pk_u = \{T_u, P_u\}$ to A_2.

Partial Private Key: On obtaining identity ID_U, S checks ID_U and ID_O equality. If equality is found, S aborts the game, else S checks L_k and returns $\{s_u, T_u\}$ to A_2.

Public Key Replacement: When query is received for user identity ID_U and $pk_u^{'} = \{T_U^{'}, P_U^{'}\}$, S looks up L_k and replaces $pk_u = \{T_U, P_U\}$ with $pk_u^{'}$. *Secret Value*: When query is received for user identity ID_U, S looks up L_K and returns x_u to A_2.

Tag-Gen: When query is received for user identity ID_U and message m_i with its identity id_i, S raises query H with id_i it returns a tuple.

1. If $c_i = 0$, S computes Tag S_i and returns $\{m_i, \text{id}_i, S_i\}$ to A_1.
2. Else if $c_i = 1$, S aborts the game.

Finally, A_2 outputs a signature $\sigma^{'}$ for ID on a message $m^{'}$. If signature can pass the proof verification equation, then we show the probability that A_2 successfully forges a signature as follows:

- M_1: S does not abort set partial private key algorithms
- M_2: A_2 successfully forges a signature σ on m for ID_O.

- M_3: The forged sign σ allows $\mathrm{ID}_O = \mathrm{ID}_O^{'}$.

From the test runs, we know

$$Pr[M_1] \geq \left(1 - \frac{q_{H_1}}{q}\right)^{q_{H_1}}$$

$$Pr[M_2|M_1] \geq \epsilon$$

$$Pr[M_3|M_1 \wedge M_2] \geq \frac{q_s}{q}$$

where A_2 can make max hash queries and sign queries. Then, the probability that S breaks the CDH problem is only:

$$Pr[M_1 \wedge M_2 \wedge M_3] = Pr[M_1].Pr[M_2|M_1].Pr[M_3|M_1 \wedge M_2] \geq \left(1 - \frac{q_{H_1}}{q}\right)^{q_{H_1}} \epsilon \frac{q_s}{q}.$$

Since ϵ is non-negligible, A_2 cannot break the CDH problem. Hence, the proposed ECLPPPA scheme is secure against Type-II adversary.

Theorem 4 (Privacy Preserving while Data Auditing) *In the ECLPPPA, TPA cannot get any information about Users data during auditing process.*

Proof In this game, A_3 tries to gain access of the system. For processing S_i, the cloud utilizes tags. Since, data blocks are masked by the user with its private key, it is computationally not possible to decipher message. (According to Theorem I). Further, users use a random number "b" and this is the information only cloud can access. So without that random number, TPA cannot decipher message. Thus, TPA cannot get any data about the data during verification. Hence, the proposed ECLPPPA is secure against Type-III adversary.

During auditing procedure, TPA knows only these information:

$$\begin{aligned} \text{Challenge} &= [\mathrm{ID}_O, a, J] \\ \sigma &= \sum_{i \in I} w_i \sigma_{m_i'} \\ \mu &= \sum_{j \in J} (w_j m_j + b \cdot H_1(\mathrm{ID}_j, B, P_{\mathrm{Pub}})) \mathrm{mod} \cdot q \end{aligned}$$

As hash functions are secure, TPA cannot derive anything about data blocks from proof.

And from μ value of b cannot be guessed as it a unknown number.

Ultimately, TPA cannot derive any information from σ (Table 1).

Table 1 Comparison of security requirements

Schemes	Correctness	Public auditing	Privacy preserving
Karati et al. [22]	Yes	No	No
Our's	Yes	Yes	Yes

5 Performance Analysis

In this section, we measure the computation cost for signature generation, proof generation and verification. Then compare the experimental results of the our scheme with Karati et al. [22]. The experimental results carried on Intel Core i7 with 2.59 GHz clock speed using charm-crypto library in Python. In experiments, Karati et al. [22] scheme assume that G_1 is a additive group and G_2 is a multiplicative group, whereas in our scheme use a single additive group G_1.

The computational cost for the generation of signatures is linear to the number of blocks. Figure 2 illustrates the comparison of computational cost for signature generation between the proposed scheme and Karati et al. [22]. From Fig. 2, we can notice that our scheme takes less time than[22] because our scheme utilized the short signatures.

In the computation cost for generating proof to provide the proof of data storage, CSP computes proof for the challenged blocks. Figure 3 shows the computational cost of CSP in the proposed scheme and Karati et al. [22] scheme for different number challenged blocks. From Fig. 3, we can notice that our scheme takes less time than existing scheme[22]. Because, our scheme verifies the only selective blocks instead of all blocks.

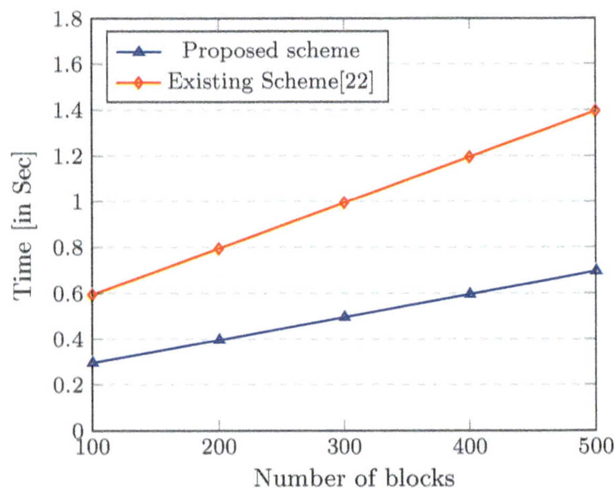

Fig. 2 Signature generation cost

Fig. 3 Proof generation cost

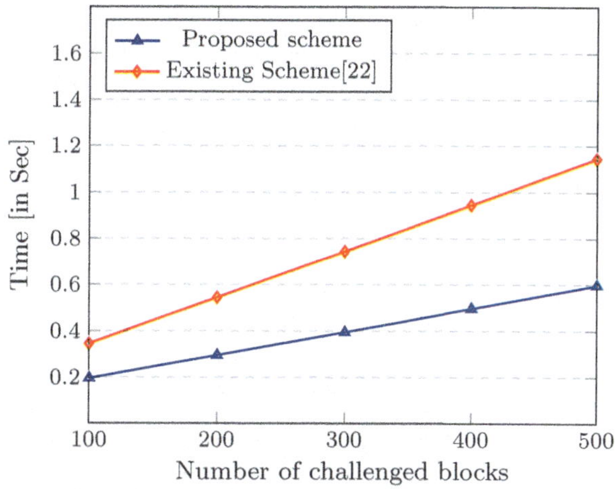

Fig. 4 Verification cost

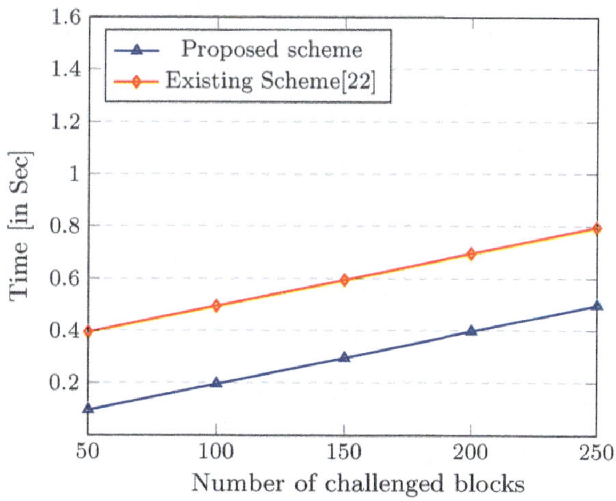

Figure 4 shows the computational cost for verification of proof. From Fig. 4, we can see that computation cost of our scheme for verification is less than Karati et al. [22] scheme because our scheme requires one pairing operation and only few blocks to verify the authentication of data.

6 Conclusion

In this paper, we proposed an efficient certificateless privacy-preserving public auditing scheme for cloud-based IIoT. Unlike previous schemes, ECLPPPA scheme requires less computational cost for signature generation and verification. Further,

ECLPPPA scheme utilized public verifier instead of data user to verify the integrity of data and reduced the overhead of data user. ECLPPPA also achieved the privacy of data from public auditor during verification. Security of ECLPPPA is proved against Type-I, Type-II and Type-III adversaries. The performance analysis shows that the our scheme is efficient than the previous scheme and suitable for the cloud-based IIoT.

References

1. Iera A, Atzori L, Morabito G (2010) The internet of things: a survey. Comput. Netw 54(15):2787–2805
2. Wachsmann C, Sadeghi TAR, Waidner M (2015) Security and privacy challenges in industrial internet of things. In: Annual Design Automation Conference. Springer, pp 1–6
3. wan J et al (2016) Software-defined industrial internet of things in the context of industry 4.0. IEEE Sens 16
4. Truong HL, Dustdar S (2015) Principles for engineering iot cloud systems. IEEE Cloud Comput 02(02):68–76
5. D'Oro S, Restuccia F, Melodia T (2018) Securing the internet of things in the age of machine learning and software-defined networking. Internet Things J 05(06):4829–4842
6. Elgamal T (1985) A public key cryptosystem and a signature scheme based on discrete logarithms. IEEE Trans Inf Theory 31(4):469–472
7. Shamir A (1984) Identity-based cryptosystems and signature schemes. In: Advances in Cryptology. Springer, pp 47–53
8. Al-Riyami SS, Paterson KG (2003) Certificateless public key cryptography. In: International conference on the theory and application of cryptology and information security. Springer, pp 452–473
9. Choi KY, Park JH, Lee DH (2011) A new provably secure certificateless short signature scheme. Comput Math Appl 61(7):1760–1768
10. Tso R, Yi X, Huang X (2008) Efficient and short certificateless signature. In: International conference on cryptology and network security. Springer, pp 64–79
11. Xu J, Zhang Z, Wong DS, Feng D (2006) Certificateless public-key signature: security model and efficient construction. In: International conference on applied cryptography, network, security. Springer, pp 293–308
12. Susilo W, Wong DS, Huang X, Mu Y, Wu W (2007) Certificateless signature revisited, in information security and privacy. In: Information security and privacy. Springer, pp 308–322
13. Shim K (2018) A new certificateless signature scheme provably secure in the standard model. Syst J 99:1–10
14. Tsai J-L (2017) A new efficient certificateless short signature scheme using bilinear pairings. IEEE Syst J 11(04):2395–2402
15. Park JH, Choi KY, Lee DH (2011) A new provably secure certificateless short signature scheme. Comput Math Appl 61(07):1760–1768
16. Wang G, Xia Q, Yu Y, Mu Y, Yang B (2012) Improved certificateless signature scheme provably secure in the standard model. IET Inf Secur 6(02):102–110
17. Yuan Y, Wang C (2014) Certificateless signature scheme with security enhanced in the standard model. Inf Process Lett 114(09):492–493
18. He D, Chen J, Zhang R (2012) An efficient and provably-secure certificateless signature scheme without bilinear pairings. Int J Commun Syst 25(11):1432–1442
19. Tian M, Huang L (2013) Cryptanalysis of a certificateless signature scheme without pairings. Int J Commun Syst 26(11):1375–1381

20. Gong P, Li P (2015) Further improvement of a certificateless signature scheme without pairing. Int J Commun Syst 27(10):2083–2091
21. Islam SH, Karati A, Biswas G (2018) A pairing-free and provably secure certificateless signature scheme. Inf Sci 450:378–391
22. Karati A, Hafizul Islam SK, Karuppiah M (2018) Provably secure and lightweight certificateless signature scheme for iiot environments. IEEE Trans Ind Inform 14(8):3701–3711
23. Mei Q, Xiong H, Zhao Y (2019) Efficient and provably secure certificateless parallel key-insulated signature without pairing for IIoT environments. Syst J
24. Zeadally S, He D, Wu L (2018) Certifcateless public auditing scheme for cloud-assisted wireless body area networks. Syst J 99:64–73

Classification of Virtual Machine Consolidation Techniques: A Survey

Saloni Sureja and Tarannum Bloch

Abstract Cloud computing is an on-demand technology for several IT infrastructures due to many of its aspects. One such aspect is virtualization, which is used for providing platform to deal with resource utilization, workloads, etc. Large data centers emit enormous amount of energy, and virtual machine consolidation is an effective technique to reduce the carbon footprints of data centers. VM consolidation accommodates virtual machines into a less number of physical machines and puts an underutilized server to hibernation mode. This paper contributes novel taxonomy of virtual machine consolidation techniques. We have derived the comparison matrices which represents the comparative analysis of performance matrix, issues resolved and mathematical models used by different VM consolidation techniques for making efficient consolidation decisions. This survey will also be helpful to the researchers intending to work for the development of decision support system for energy consumption minimization and to achieve good quality of service.

Keywords Cloud · Virtualization · Consolidation · Energy consumption · Resource management

1 Introduction

Nowadays, the prime concern for the cloud user is to store business data securely and safely in a more scalable environment with higher level of security. In current era, users are much concerned about reduction of IT infrastructure costs and want their data to be accessed from anywhere and anytime. The cloud technology is an illustration for the Internet to achieve aforesaid requirements of any business model. Cloud computing is an Internet-based model which enables sharing of resources,

---

S. Sureja (✉)
Department of Computer Engineering, Marwadi Education Foundation, Rajkot, Gujarat, India
e-mail: saloni.sureja16499@marwadieducation.edu.in

T. Bloch
Department of Information Technology, Marwadi University, Rajkot, Gujarat, India
e-mail: tarannum.bloch@marwadieducation.edu.in

S. M. Thampi et al. (eds.), *Applied Soft Computing and Communication Networks*, Lecture Notes in Networks and Systems 125,
https://doi.org/10.1007/978-981-15-3852-0_11

data to the computers, other devices based on the demands of users. There are various definitions for cloud computing according to Shaw [1] "cloud computing is a popular service provisioning model which offers users access to an on-demand resource accessible over the Internet and charged on a pay-as-you-go basis." As per our understanding, "cloud is a multi-user system where resources are shared among the diverse application in dispense environment." The cloud service model provides users with services for deploying their applications. Services can be infrastructure as a service (IaaS), platform as a service (PaaS) and software as a service (SaaS), which is provided by cloud service providers. The cloud deployment model represents exactly the kind of environment user wants to work with cloud. Deployment models can be public, private and hybrid. Apart from this, there are many other aspects where cloud computing comes into picture such as data communication and mobile computing [2] for extensive computations. To fulfill the demands of dense population of users in wireless environment, technologies such as software-defined network (SDN) and cognitive radio networks (CRNs) [3, 4] are coming up with idea of edge computing. SDN simplifies network management and CRN for better utilization of wireless spectrum. However, the big challenge is to perform computation in low-power handheld end user devices and to extract, store and manage this data over cloud server.

The paper is organized as follows: Sect. 1 describes VM consolidation; Sect. 2 describes the state-of-the-art techniques. Section 3 depicts a novel taxonomy of virtual machine consolidation techniques. Section 4 describes performance assessment matrix and issues resolved by existing approaches. Section 5 provides comparative study of the available approaches, and Sect. 6 provides concluding remarks and the future scope.

1.1 Virtual Machine Consolidation

Whenever cloud computing comes in context, virtualization is always in consideration. It suggests the virtual version of the constituent elements and provides the proper utilization of resources by running multiple workloads on the same server. Virtual machines are created to hold the emulation property and to avail the functionality of a physical machine, to accommodate multiple users on single servers.

Despite the benefits achieved from advancement in virtualization, cloud relies on large data centers so large amount of energy consumption has been observed as a critical issue in the data centers. Virtual machine consolidation minimizes the energy consumption by accommodating more instances of running virtual machines into a less number of active servers and putting the underutilized servers in hibernation mode which saves a greater amount of energy. Figure 1 depicts the virtual machine consolidation, and there are three hosts: host A, host B and host C. As shown, host B consists of only one virtual machine due to which the physical machine resources are not utilized properly and energy is being wasted. Thus to minimize the energy consumption of host B, virtual machine must be migrated to host A as its resources

Fig. 1 VM consolidation

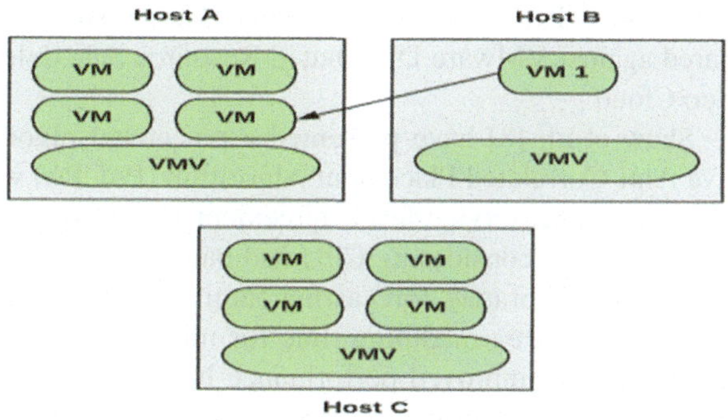

are not fully utilized, thus depicting the idea of consolidation. The virtual machine cannot be accommodated to host C as it is already utilizing all the resources.

2 State of the Art

Mohiuddin and Almogren [5] have introduced virtual machine consolidation model (2018) by inferring virtual machine migration technique. In order to enhance green computing, this technique consolidates the virtual machines according to their workloads to reduce the number of physical machines to reduce the energy consumptions. A classification-based heuristic method was adopted to ensure load balancing at the time of virtual machine allocation. Further authors have addressed various issues such as the use of hazardous materials, energy consumptions, wastage of resources and reshuffling of virtual machines. The proposed method is compared with simulated annealing, genetic algorithm and reduced energy consumption up to 15%.

Sotiriadis et al. [6] discussed the virtual machine scheduling algorithm (2017) for minimizing performance degradation and maximizing CPU utilization by considering the usages of currently running virtual machine resources by analyzing previous utilization levels for scheduling virtual machines. For prediction of resource usages, various machine learning models were trained in order to build a monitoring system that gathers data usages of the resources. The algorithm allows processing of data based on the time frame window for defining the actual behavior of physical machines and virtual machines. The solution is compared with traditional schedulers in OpenStack, minimizes up to 19% performance degradation and enhances CPU utilization by 2%.

Guo et al. [7] have introduced self-adaptive virtual machine allocation and migration algorithms (2019) for energy-efficient scheduling which aims at reduction energy of consumption of all the physical machines and to enhance the resource utilization, allocation and migration probability functions for optimization purpose, thus ensuring a good quality of service. The authors have not considered load balancing during experiment. The proposed algorithm reduces up to 30% of energy consumption compared against VMware Distributed Resource Scheduler (DRS) and 10–20% against Eco Cloud.

Shaw et al. [8] have presented a placement algorithm (2018), namely Predictive Anti-Correlated Placement Algorithm (PACPA) which aims at reducing energy and increasing service-level agreement (SLA) over some of the virtual machine placements by considering CPU and bandwidth resource consumptions. It resolved issues like accurately forecasting future resource demands in a dynamic environment to improve scheduling time for migrations and also for placement of virtual machines and improved performance based on recent approaches. For forecasting autoregressive integrated moving average (ARIMA) and artificial neural network (ANN), methodologies have been used and compared to the performance of ANN and ARIMA in order to decide the superiority of the models. The PACPA reduced up to 18% energy and achieved 47% SLA when compared with best fit, first fit and power-aware best-fit decreasing (PABFD). In the future, the research can be extended by regression analysis of the proposed system on a real platform.

Farhadian et al. [9] discussed the imperialist competitive and genetic algorithm (2019) which applies evolutionary methods by employing different virtualization techniques. This method aims at providing optimum virtual machine allocation to physical machines for improvement in energy consumption. The algorithm includes five steps based on genetic algorithm principles: (1) Initial empires will be formed, (2) assimilation policy is applied to move colonies nearer to imperialists, (3) crossover is applied to get the optimum answers, (4) revolution operators are applied to check diversity, and (5) cost function is calculated. Imperialist competitive and genetic algorithm improves around 19.39% of energy consumption when compared with power-aware best-fit decreasing.

Wang et al. [10] have designed consolidation mechanism for server reliability and optimization of resource allocation. The authors have considered host power consumptions and variability in temperatures to gain trade-offs between energy saving and SLAV. The mechanism adopts Markov model to predict future CPU usages for migration purpose.

Shaw et al. [11] have introduced a reinforcement learning technique (2017) which presents the virtual machine consolidation model for improving resource management by optimizing the virtual machine distribution. The model proposes advance reinforcement learning consolidation agent, capable of enhancing the efficiency and quality of service. The model consists of (1) state–action space and (2) advance reinforcement learning consolidation agent learning model. The Advanced Reinforcement Learning Consolidation Agent (ARLCA) is compared with Local Regression Minimum Migration Time (LRMMT) policy and observed that ARLCA outperforms LRMMT by achieving a significant 76.2% decrease in SLA violation.

Alsadie et al. [1] presented the two new approaches (2018): Multiple Resource-based VM Selection (MRVMS) and Lowest Interdependence Factor Exponent Multiple Resources Predictive (LIFE-MP) for addressing the limitations of existing approaches for VM consolidation. Multiple Resource-based VM Selection selects VM with high CPU requirements for reducing the workloads, thus minimizing the migration cost. Lowest Interdependence Factor Exponent Multiple Resources Predictive (LIFE-MP) method selects physical machines with the lowest correlation coefficient value among already running virtual machines to reduce the performance degradation by migrating the virtual machines. The approach minimizes the number of migrations, number of SLA violations and power consumptions of the PMs. Lowest Interdependence Factor Exponent Multiple Resources Predictive achieved a 29.02% reduction in energy consumption, service-level agreement violations by 32.68% and number of migrations up to 66.09% when compared with power-aware best-fit decreasing scheme.

Farahnakian et al. [12] introduced utilization-prediction-aware virtual machine consolidation approach (2016) to address the problem of unnecessary virtual machine migrations. The approach takes into account: (1) For identifying service-level agreement violations, both memory and CPU utilizations are taken into consideration, (2) it uses two prediction models: K-nearest neighbor and linear for future utilization, and (3) investigate the performance of UP-VMC based on three virtual machine selection policies which are (1) minimum migration time (MMT), (2) maximum load (Max L) and (3) minimum load (Min L) in the context of energy consumption, number of service-level agreement violations (SLAV) and number of migrations. The research can be further extended for improvement of scalability of the proposed architecture, by considering network resource utilization and network traffic for virtual machine placement. The proposed algorithm reduces energy consumption and SLAV when compared with modified best-fit decreasing and modified first-fit decreasing.

Haghshenas et al. [13] presented a multi-agent machine learning-based approach (2019) that tackles energy efficiency and variability in performance. The approach is a machine learning approach which migrates the virtual machine by choosing the acceptable power mode and frequency of each host at the time of execution. The approach is capable of detecting performance variability and thus migrates all the virtual machines from the degraded physical machines. It reduces up to 15% of energy consumption in comparison with the fast best-fit decreasing (FBFD) and power-aware best-fit decreasing (PABFD). It achieved up to 86% reduction in number of migrations and host power mode transition by up to 90% and thus improved scalability than other existing approaches. In the future, the research can be extended for heterogeneous/hybrid CPU data centers for parallel processing of applications.

Zhou et al. [14] proposed a novel adaptive energy-aware algorithm (2018). Their contribution includes: (1) threshold-based framework with four classes to deal with the dynamic and unpredictable workload, (2) VM selection methods, virtual machine migration, CPU-intensive task and I/O-intensive task by considering CPU and memory utilization for making the decisions, (3) virtual machine deployment policy to maximize energy efficiency and (4) the adaptive energy-aware algorithm is being evaluated using real-world loads and results in reducing energy consumption and SLA violations up to some extent. In the future, the focus can be achieved by studying improvements in energy efficiency and reduction of operation cost.

Alharbi et al. [15] presented an approach (2019) for dynamically placing the virtual machines to physical machines in the data center. It considers the virtual machine placement problem as a constrained combinatorial optimization problem with objective of minimizing energy based on constraints by making use of PM-VM profiles. Secondly, an ant colony system is used with new heuristics related to the usage of PMs and capacity for gaining energy-efficient solutions. Comparison with two ant colony system (ACS) methods and first-fit decreasing (FFD) algorithm, ACS outperforms the FFD with respect to energy performance by achieving 10% improvement in energy efficiency and improves scalability for larger problems.

Sharma et al. [16] introduced failure-aware virtual machine consolidation mechanism (2018) to address the problem of reliability by considering failure events and risky rates of physical resources before consolidation of virtual machines. A failure prediction approach is proposed in order to produce two fault tolerance mechanisms that are (1) VM migration: migrating the running VMs to more reliable resources and (2) VM check pointing: In order to save the state of the virtual machine, check pointing is used. Here, checkpoints are created based on failure prediction results and based on that evaluate with VM migration. Further on combining the reactive and proactive with failure-aware mechanism, it was observed that it achieved about 34% reduced energy, 12% improved reliability and 14% decrement in occurrences of failure. In the future, new methods can be designed with correlated failures to achieve fault tolerance in an efficient way and the proposed method will be tested with correlated failures.

Cao et al. [17] proposed a distributed virtual machine dynamic consolidation framework (2019) for managing dynamic consolidation efficiently. A fine granular design is being introduced by considering the capacities of each switch level to coordinate network and communication topology. In order to minimize energy utilization and performance constraints caused due to communication traffic, the framework adopts multi-objective-based genetic algorithm for VM allocation. The approach outperforms the previous approaches when compared to energy consumption, bottleneck ratio and service-level agreement violation ratio. The research can be extended further on investigating the proposed technique with routing protocol for reducing network constraints, and a new approach can be designed for greater performance with the dynamic routing protocol.

Haghighi et al. [18] applied a virtualization technique which introduced k-means clustering and dynamic consolidation method improved by the micro-genetic algorithm (KMGA) (2019). K-means clustering and micro-genetic algorithm (KMGA) aims at providing the proper balance between the negotiated service-level agreements and energy consumption of the servers. In comparison with particle swarm optimization and the genetic algorithm, it minimizes the number of virtual machine migration, energy consumption and make span. In the future, the research can be extended by considering fuzzy logic for clustering, neural network model to predict the overloaded hosts and dynamic scheduling algorithms to efficiently map the tasks.

3 Virtual Machine Consolidation Classification Techniques

Virtual machine (VM) consolidation is an efficient way to achieve energy savings by running virtual machines into less no. of possible servers and putting the underutilized server in hibernation mode. VM consolidation techniques contribute advantages like load balancing, energy efficiency, resource utilization and management. Figure 2 presents a hierarchical diagram of classification.

3.1 *Optimization Method*

One of the approaches for solving VM consolidation problem is optimization technique. The optimization technique basically searches for optimum solutions for the problems. Mainly, there are two types of approach: exact and approximate approaches. Exact methods find the optimal solutions but are time-consuming as it is an NP-hard problem and the problem will increase exponentially with the increase in problem size. To solve this problem, heuristic and meta-heuristic methods are used

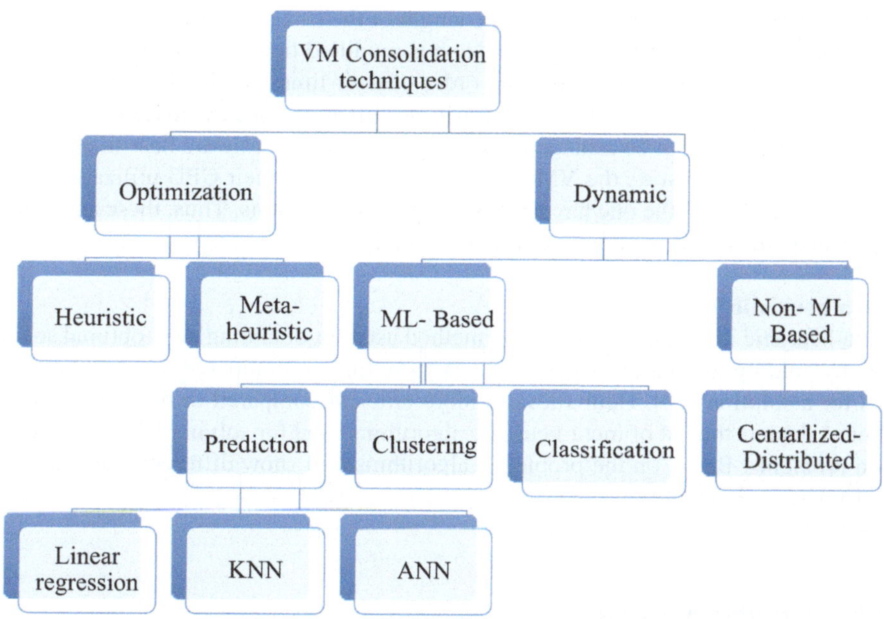

Fig. 2 Classification of VM consolidation techniques

Table 1 Meta-heuristic approaches to solve VM consolidation

Meta-heuristic	Used by
Genetic algorithm	Farhadian et al. [9]
Micro-genetic algorithm	Haghighi et al. [18]
Hybrid optimization	Haghighi et al. [18]
Ant colony system	Alharbi et al. [15]

[19], whereas approximate approach does not ensure you with a best solution but it can provide with a near-optimal solution to a problem within the polynomial time.

Heuristic Method
Heuristic methods are problem-dependent methods generally used to get near-optimal solutions efficient enough to achieve near-immediate goals. It does not guarantee to provide an optimal solution but near-optimal solutions. Heuristics are an efficient method for optimal allocation of virtual machines [19]. Heuristics are used for migration as well as allocation [8]. Some heuristics defined are first fit [8]. The algorithm first sorts the tasks in decreasing order of their processing time length and allocates the tasks to machines based on acceptance of a task according to this sequence. Another algorithm is bucket first fit (BFF). The algorithm divides the tasks into different buckets based on the length of the tasks and assigns the physical machines on the bucket as per first fit. Next is modified first-fit decreasing (MFFD) which arranges the jobs in decreasing order of their time lengths. If both the time lengths are same, the one with first starting time will be arranged with respect to each other; otherwise, it can be arranged randomly. Next is power-aware best-fit decreasing (PABFD) which sorts the VMs in decreasing order of their CPU utilization and allocates the VMs to the one having less energy consumptions. Thus, these are some of the heuristics used in gaining optimal solutions.

Meta-heuristic Method
Meta-heuristic is another optimization method used for obtaining near-optimal solution by guiding the search space but takes more time as compared to quick heuristics for a solution [10]. Heuristics are more efficient compared to meta-heuristics. Table 1 depicts the list of meta-heuristic algorithms used for solving VM consolidation problems. Based on the problems, algorithms will show different benefits and performances in nature.

3.2 Dynamic Method

In dynamic consolidation, the basic idea is to place the virtual machine to the appropriate physical machines and switching the idle PMs into hibernation mode in order to reduce effective energy consumptions in data centers for optimizing performance degradation, ensure load balancing, maintaining SLA, reliability, etc., promoting green computing [5]. This technique is widely used for consolidation. The approach would be capable of detecting performance variability and thus migrates all the virtual machines from the degraded physical machines [15]. For dynamically consolidating the virtual machines, there are two approaches: ML-based approach and non-ML-based approach.

Machine Learning-Based Method

Nowadays, there is rapid growth in machine learning techniques for each and every technology. Cloud computing also uses ML-based techniques in order to achieve overall efficiency in cloud data centers. Rachael Shaw et al. mentioned the prediction techniques for accurately predicting the future resource demands for optimum resource allocation and resource managing [11]. Yogesh Sharma et al. discussed the failure prediction techniques for reliability factor of resources by considering occurrences of failures before the consolidation of VMs, thus improving the reliability factor [16]. Several other approaches have been presented like clustering and classification. Mainly, clustering is adapted for mapping the tasks that are mapping the relevant task from a cluster to relevant machines [18]. Further prediction can be classified into artificial neural network (ANN)-based prediction and autoregressive integrated moving average (ARIMA) [19].

Prediction

Nowadays, the major concern in cloud computing is resource management and provisioning in order to predict the future resource requirements for the sake of minimizing the VM migration and reducing energy consumptions in the data centers. Fahimeh Farahnakian et al. discussed a VM consolidation approach that takes current and future resource demands into consideration. It uses a regression-based model for the prediction that is a linear and K-nearest neighbor. In addition, K-nearest neighbor is used to estimate the future utilization of VMs and PMs from historical datasets [12]. Further several approaches are adopted for prediction purpose. Such approaches are artificial neural network, linear regression and K-nearest neighbor (KNN).

Linear Regression

According to Fahimeh Farahnakian et al., linear model is used to predict the resource consumptions of the CPU [12]. A linear function is defined which estimates the linear relationship between the present and future resource utilizations. This method estimates the line that best fits the graph, the one that gives the minimum distance between the predicted output and the observed output. Thus, this method can be considered as an efficient method for predicting the resource utilization and thus reducing energy consumptions and number of VM migrations.

K-Nearest Neighbor

According to Farahnakian et al. [12], K-nearest neighbor (KNN) is a nonparametric method used for regression. It considers the local average of datasets, and based on that it estimates the future resource utilization. By locality, it means k-nearest samples to the new samples and the one with the best value is chosen. Leave-one-out is a cross-validation method which makes prediction accurate by performing summation of all the squared residuals for each value of k, and based on these values it selects the k with a minimum of the residuals.

Artificial Neural Network

Artificial neural network (ANN) is a very well-known technique used in machine learning for various purposes. It is used for forecasting future resource demands. According to Shaw et al. [19], a feed-forward neural network is proposed which consists of the input layer, hidden layer and the output layer. Here, CPU utilization is given as an input vector. Further, these values are given to the hidden layer, along with the weights. Further, the network calculates sum of these weighted signals and gives the output, respectively. An activation function is used at the output to limit the range of signal between 0 and 1. After the signal has been generated, it generates a predicted output for estimating future resource demands for efficient VM consolidation.

Clustering

It is one of the machine learning techniques where objects with similar properties are grouped into one cluster by calculating the Euclidean distance based on the mean values of each cluster. As per Haghighi et al. [18], different cloud resources are classified into the cluster based on their diverse nature as the k cluster number. Based on this k, the VMs and the tasks would be grouped by considering their processing power and the length of time. Further, the suitable cluster of tasks would be mapped to their suitable VMs. It follows the first come first serve (FCFS) fashion for allocation of tasks to the resources.

Classification

Classification is one of the very well-known techniques of machine learning. Mohiuddin et al. [5] have used classification to ensure load balancing at the time of allocation. The physical resources are classified based on their resource capacities like CPU, bandwidth and memory. Physical machines are divided into 4 classes, i.e., class 1, class 2, class 3 and class 4 where class 1 comprises higher capabilities and class 4 as the minimum amount of resources. The classifier will check for the available classes in the physical machines based on the VM request for the resources and accordingly map the VM. Thus before allocation, the classifier will read the request and map them accordingly. Furthermore Zhou et al. [14] divided the hosts into 4 different classes of the less loaded host, little loaded host, normally loaded host, and overloaded host based on the CPU utilization of each host. The classification is done for the migration of VMs accordingly based on workloads. It uses the threshold framework for classifying the host in order to reduce service-level agreement violations and reduce energy consumptions.

Non-ML-Based Method
Non-ML-based methods are methods rather than predictions, clustering, etc. It does not use learning-based methods for VM consolidation. The non-ML-based method has focus on centralized-distributed technique.

Centralized-Distributed Approach
Haghshenas et al. [13] have used the approach consisting of power mode selection agents distributed to each host, combined with the centralized unit for VM migration decisions. These power mode selection units will select the most relevant mode for the hosts and improve the deal between service-level agreement violations and energy consumptions. Further to content these power modes, certain migrations will be performed by a migration unit. Also to resolve the over-utilization problem, migration unit detects the over-utilized hosts by the concept of agent, state and reward and then migrates the VM.

4 Performance Access Matrix

The VM consolidation should be able to migrate the VM from underutilized PMs to well-fitted PMs. There are various factors on which consolidation depends:

Energy consumption
It is the consumption or utilization of energy or power consumed by cloud servers during the processing of PMs.

Throughput
It is the rate at which the service request is being processed. **Bandwidth**: It is the maximum amount of data transferred over a network in a given amount of time.

The number of migrations
It is the migration of VMs required for the proper utilization of PMs.

Outage/Downtime
It is the duration when the server fails to perform during the consolidation process.

4.1 Issues Resolved

The various issues have been resolved during the consolidation process.

Power saving
The saving of power consumption by putting idle servers off or in sleep mode.

Resource utilization
Utilization of resources like CPU, RAM and many more as per requirements.

Performance degradation
It is the breach of discussed service-level agreement contract between users and brokers due to over-utilization of resources, server failure, etc.

Service-level agreement
A performance conversed between the customer and service providers for observing of services.

ML for prediction
Predicting the resource demands based on current and future resources for efficient management of resources.

Reliability
Providing consistent services for client conveniences.

5 Comparative Study

Different VM consolidation techniques have focused on various parameters. Table 2 gives comparative study of the performance matrix of the existing approaches. Table 3 gives brief issues resolved by the existing approaches, and Table 4 depicts the mathematical model and the tools used for implementation.

6 Conclusion

For cloud computing, virtual machine consolidation is an efficient technique to gain insights regarding energy consumptions, resource utilization and SLA violations. Recent studies have been done to avert reliability issues, performance degradation, etc. This paper presents a classification of VM consolidation techniques based on adopted methodology of existing VM consolidation techniques. Comparison tables show that ML approach resolves most of the discussed issues. In the future, decision support system comprising monitoring capabilities for VM consolidation using machine learning can be developed to reduce energy consumption in order to promote green computing.

Table 2 Comparative performance matrix of existing approaches

Topic/ref	Energy consumption	Throughput	Bandwidth	No. of migrations	Outage
[5]	✓	×	×	×	×
[7]	×	✓	×	×	×
[8]	✓	✓	×	×	×
[9]	×	×	✓	×	×
[10]	✓	×	×	×	×
[11]	✓	×	×	✓	✓
[19]	✓	×	×	✓	×
[1]	✓	×	×	×	×
[12]	✓	×	×	✓	×
[13]	✓	×	×	✓	×
[14]	✓	×	×	✓	×
[15]	✓	×	×	✓	×
[16]	✓	×	×	✓	×
[17]	✓	×	×	×	×
[20]	✓	×	×	×	×
[18]	✓	×	×	×	×
[21]	✓	×	×	✓	×

Table 3 Issues resolved by existing approaches

Topic/ref	Power saving	Resource utilization	Performance degradation	SLA	ML for prediction	Resource management	Reliability
[5]	✓	✓	×	×	×	×	×
[7]	×	×	✓	×	×	×	×
[8]	✓	×	×	×	×	×	×
[9]	×	×	✓	✓	×	×	×
[10]	×	×	×	×	×	×	×
[11]	✓	✓	×	✓	×	×	×
[19]	✓	×	✓	✓	×	×	×
[1]	✓	×	✓	✓	✓	✓	×
[12]	×	×	×	✓	×	✓	×
[13]	×	×	×	✓	✓	×	×
[14]	×	×	×	✓	✓	✓	×
[15]	✓	✓	×	✓	×	×	×
[16]	×	×	×	✓	×	×	✓
[17]	×	×	×	×	×	×	×
[20]	×	×	×	×	×	×	✓
[18]	×	×	×	✓	×	×	×
[21]	×	×	×	✓	×	×	×

Table 4 Mathematical model and tools used for implementation

References	The methodology adopted by existing solutions used	Tools used
[5]	• Workload-aware virtual machine consolidation method (WAVMC) model for VM migration • Classification for VM placement based on workload • VM migration for resource management using workload	MATLAB
[6]	• VM scheduling to optimize performance • The machine learning model for prediction	OpenStack
[7]	• Self-adaptive VM consolidation (SAVE) method to reduce energy consumption	CloudSim
[8]	• Predictive Anti-Correlated Virtual Machine Placement Algorithm (PACPA) to reduce energy and SLA violation • Prediction models such as artificial neural network (ANN), autoregressive integrated moving average(ARIMA), moving average (MA) and random walk (RW) for resource demands	CloudSim
[9]	• Imperialist competitive and genetic algorithm (ICGA) method for optimum VM allocation	CloudSim
[10]	• Thermal-aware VM consolidation Markov model for optimum allocation and reliability assurance	CloudSim
[11]	• Advanced Reinforcement Learning Consolidation Agent (ARLCA) learning model for reducing energy consumption and SLA violations	CloudSim
[1]	• Multiple Resource-based VM Selection (MRVMS) for VM selection • Lowest Interdependence Factor Exponential Multiple Resource Predictive (LIFE-MP) for VM placement	CloudSim
[12]	• Regression-based model (UP-VMC) for future CPU and memory utilization of VMS and PMs	CloudSim
[13]	• Multi-agent machine learning-based approach (MAGNETIC) for energy efficiency • Host power model to estimate host power • VM migration energy model	CloudSim
[14]	• Adaptive three threshold frameworks Ta, Tb, Tc($0 \leqq Ta < Tb < Tc \leqq 1$)	CloudSim
[15]	• Ant colony system with an embedded heuristic for an energy-efficient solution	CloudSim
[16]	• A mathematical model for reliability, energy consumption and finishing time • Failure-aware energy-efficient VM consolidation technique for energy wastage	CloudSim
[17]	• Virtual machine dynamic consolidation (VMDC) framework for dynamic consolidation management	CloudSim
[20]	• Bin-packing heuristic for VM placement	CloudSim
[18]	• K-means clustering and micro-genetic algorithm (KMGA) to reduce energy consumption	CloudSim

References

1. Alsadie D, Tari Z, Alzahrani EJ, Alshammari A (2018) LIFE-MP: online virtual machine consolidation with multiple resource usages in cloud environments. In: Hacid H, Cellary W, Wang H, Paik HY, Zhou R (eds) Web Information Systems Engineering–WISE 2018. Lecture notes in computer science, vol 11234, WISE 2018. Springer, Cham
2. Dutta N, Misra IS (2014) Multilayer hierarchical model for mobility management in IPv6: a mathematical exploration. Wirel Pers Commun 78(2):1413–1439, Springer
3. Dutta N, Sarma HKD, Polkowski Z (2018) Cluster based routing in cognitive radio Adhoc networks: reconnoitering SINR and ETT impact on clustering. Com Com 115:10–20, Elsevier
4. Dutta N, Sarma HKD (2017) A probability based stable routing for cognitive radio Adhoc networks. Wire Net 23(1):65–78, Springer
5. Mohiuddin, Almogren A (2018) Workload-aware VM consolidation method in edge/cloud computing for IoT applications. J Parall Distrib Comput 123:204–214
6. Sotiriadis S, Bessis N, Buyya R (2018) Self-managed virtual machine scheduling in Cloud systems. Inf Sci 433–434:381–400
7. Guo W, Kuang P, Jiang Y, Xu X, Tian W (2019) SAVE: self-adaptive consolidation of virtual machines for energy efficiency of CPU-intensive applications in the cloud. J Supercomput 70(121):1–25
8. Shaw R, Howley E, Barrett (2019) An energy-efficient anti-correlated virtual machine placement algorithm using resource usage predictions. J Model Simul Cloud Comput Big Data 93:322–342
9. Farhadian MK, Rezazadeh J, Farahbakhsh R, Sandrasegaran K (2019) An efficient IoT cloud energy consumption based on genetic algorithm. J Dig Commun Netw
10. Wang JV, Cheng C-T, Tse CK (2019) A thermal-aware VM consolidation mechanism with outage avoidance. PractExper, Soft, pp 1–15
11. Shaw R, Howley E, Barrett E (2017) An advanced reinforcement learning approach for energy-aware virtual machine consolidation in cloud data centers. In: The 12th IEEE international conference for internet technology and secured transactions, Cambridge, UK, 2017
12. Farahnakian F, Pahikkala T, Liljeberg P, Plosila J, Trung Hieu N, Tenhunen H (2016) Energy-aware VM consolidation in cloud data centers using utilization prediction model. IEEE Trans Cloud Comput XX(X)
13. Haghshenas K, Pahlevan A, Zapater M, Mohammadi S, Atienza D (2019) MAGNETIC: Multi-Agent Machine Learning-Based Approach for Energy Efficient Dynamic Consolidation in Data Centers. IEEE Trans Serv Comput, pp 1–1
14. Zhou Z et al (2018) Minimizing SLA violation and power consumption in cloud data centers using adaptive energy-aware algorithms. J Future Gener Comput Syst 86:836–850
15. Alharbi F, Tian Y, Tang M, Zhang W, Peng C, Fei M (2019) An ant colony system for energy-efficient dynamic virtual machine placement in data centers. J Exp Syst Appl 120:228–238
16. Sharma Y, Si W, Sun D et al (2018) Failure-aware energy-efficient VM consolidation in cloud computing systems. Future Gener Comput Syst (2018)
17. Cao G (2019) Topology-aware multi-objective virtual machine dynamic consolidation for cloud datacenter. Elsevier. https://doi.org/10.1016/j.suscom.2019.01.004
18. Askarizade Haghighi M, Maeen M, Haghpar M (2019) An energy-efficient dynamic resource management approach based on clustering and meta-heuristic algorithms in cloud computing IaaS platforms. Int J Wirel Personal Commun 104(4):1367–1391
19. Shaw R, Howley E, Barrett E (2019) A predictive anti-correlated virtual machine placement algorithm for green cloud computing. In: The 11th IEEE international conference on Utility and Cloud Computing (UCC), Zurich, Switzerland, 2019
20. Moges F, Abebe S (2019) Energy-aware VM placement algorithms for the Open Stack Neat consolidation framework. J Cloud Comput 8(1)
21. Bloch T, Sridharan R, Prashanth C (2014) Analysis and survey of issues in live virtual machine migration interferences. Int J Adv Netw Appl (IJANA)

Enhanced Honey Bee Load Balancing in Large Heterogeneous Cloud Environments

B. Nithya, Harshit Mogalapalli, Kamalesh Khanna, and Sasmita Moharana

Abstract Recently, abundant domains such as education, healthcare, science, finance and many other applications have been popular to establish the recent contemporary content-based applications. These applications need large gathering, generation and processing of data and have all of them in a big heterogeneous system consisting of a diverse range of public/private cloud systems that are dispersed geographically. With this context, provisioning and allocation of resources become a great challenge to distribute modern systems because of unpredictable workload that change or fluctuate during their execution. To cope with this, an adaptive algorithm, namely enhanced honey bee (EHB) load balancing algorithm, is proposed. The main aim of this algorithm is to distribute the workloads of numerous network links to avoid overutilization and underutilization of resources through self-aggregation and the concept of foraging nature of honeybees. This is achieved by assigning the ready tasks to virtual machine (VM) by satisfying two yardsticks. One is that the tasks count of the current available VM should be lesser than other VMs and second is that the deviation/variation of the processing time of current VM from the meantime of processing in all VMs should not exceed the predetermined threshold. The evaluated empirical results show efficiency of the proposed EHB model through the response and processing time, deviation of load and the extent/degree of the imbalance.

1 Introduction

Cloud computing is regarded as the most powerful, developing scientific technology to provide a highly fast and secure services through the Internet. In this, all software applications and data files are held on a large number of servers and are accessible through Internet. This model has gained the attention from large business firms and the IT industry for availing instant services. The providers of cloud services abstract

B. Nithya (✉) · H. Mogalapalli · K. Khanna · S. Moharana
Department of Computer Science and Engineering, National Institute of Technology Tiruchirappalli, Tiruchirappalli 620015, Tamil Nadu, India
e-mail: nithya@nitt.edu

S. M. Thampi et al. (eds.), *Applied Soft Computing and Communication Networks*, Lecture Notes in Networks and Systems 125,
https://doi.org/10.1007/978-981-15-3852-0_12

the hardware from users by giving them VM view. From economic aspect, individual VM should execute the tasks as quickly as possible to reduce response time. In heterogeneous cloud environment, balancing the workloads of all VMs becomes an utmost critical and prime research area, as it minimizes overutilization and underutilization of VMs, facilitates better service quality and reduces execution and response time [1, 2]. In spite of the challenges such as uncertain and dynamic workload, scalability with rising and falling demands of users, ability to continuously function for longer duration, there is a necessity for a mechanism in heterogeneous distributed system, which will divide the workload equally and speed up the task's execution [3–5]. The mechanism for balancing load is examined extensively in both homogeneous and heterogeneous environments. The categorization of load balancers (LB) is illustrated in [6], and they are classified into two types: static and dynamic. Static LB needs prior information regarding the parameters of the VM a priori. So, workload has to be distributed at the beginning, and it does not account the current state. Performance is calculated when the execution starts. Once the load is allocated, it cannot be given to another host. Static LB works well on homogeneous systems, where servers have fewer fluctuations in application workloads but gives poor resource utilization. If it works for heterogeneous system, then workloads vary frequently. Dynamic LB is exceptionally good for heterogeneous systems with varying workloads by paying additional cost for managing load information. In dynamic LB, the parameters of the VM can be changed dynamically, the changes are saved and workload can be allocated accordingly. It is highly efficient and scalable. The principle objective of a LB is to enhance the speed of time of executing the applications using the cloud instances. But, workloads of those applications bring unpredictable hazards while running. This becomes a strong viewpoint to measure the system load among heterogeneous servers. Three most crucial conditions, which an effective load balancer should satisfy, are:

1. It has to differentiate workloads and reorganize to scale down the non-demanding workloads.
2. It should freely adapt to the heterogeneous environment.
3. Lastly, it has to enhance the throughput. It is the quantity of requests that the system processes for a certain amount of time.

Conventional algorithms do have limitations as because the workload varies frequently. Despite researchers have addressed that it takes little effort to schedule in conventional cloud, but still, it needs great amount of work to deal with the scheduling of jobs so as to gain load balancing. To tackle with such a dynamic environment, swarm intelligence (SI) algorithms such as ant colony optimization (ACO), particle swarm optimization (PSO), artificial bee colony (ABC) become popular in recent decades [7]. These algorithms achieved exceeding progress in heterogeneous dynamic cloud environment. But, some algorithms do have drawbacks as they overloaded many hosts and got lesser throughput [8]. The proposed EHB uses the same technique as that of the colony of bees and comes under the category of dynamic LB.

A summary depicting research work for the aforementioned algorithm has proven to have outperformed or matched few meta-heuristic algorithms. It enhances overall processing performance that reduces the task's waiting time in queue, decreasing the overall latency. This research work reveals that the performance is increased as the service varieties are increased but seize to show the growth when the system size increases. The self-aggregation (SA) algorithm forms an active clustering technique that groups essential and similar services to carry out load balancing. In the usual standard environment, the authors in [9] described SA as a network of interrelated nodes with its specification and a list containing neighbor nodes. The algorithm gets on the connectivity to the neighboring nodes to found a configuration which is optimum. The algorithm characterizes the nodes with the help of the network that needs to be rewired.

Adopting these facts, the proposed EHB algorithm distributes the coming requests such that it will give better VM utilization and speed in overall time and performance. VMs, which are in an available state, are considered for assigning the incoming tasks. To avoid traffic, it allocates task VM that has minimum load. Thus, it avoids overutilization and underutilization of resources to optimize throughput, response and hosts processing time. It prevents allocation when variation in time of processing requests exceeds value of threshold. This indicates the heavy impact of standard deviation that eventually balances the whole system. The tasks, which got allocated, memorize and inform regarding the VM's status to waiting tasks in queue, like the way bees memorize and inform other bees with their waggle dance when they find plenty of nectar source.

1.1 Contribution

The proposed enhanced honey bee (EHB) algorithm figures out how to assign the users requests to hosts of the DCs. It preserves the balance of application workloads by combining SA and bee's algorithm that is specifically targeted for a large heterogeneous and distributed cloud system. This paper contributes the following:

1. A hybrid-novel algorithm and its implementation target at efficient resource allocation. The proposed EHB model is self-organizing intelligent solution that effectively utilizes the resources in heterogeneous cloud.
2. The evaluation metrics is computed after running test cases for empirical analysis and compared with the existing algorithms for load balancing in heterogeneous cloud: RR [10], Throttled [6], ESCEL [6], ACO [11] and PSO [12].
3. Empirical evaluations of EHB is conducted via a simulation using the extension of CloudSim Simulation toolkit framework in terms of response time, processing time, load standard deviation and the degree/extent of imbalance.

The rest of the paper is organized as follows: Sect. 2 discusses about the classification of load balancing algorithms and more details of honey bee-based algorithm; Sect. 3 gives complete idea about the proposed enhanced honey bee-based load balancing algorithm. Section 4 shows the simulation results, and finally, the last section of the paper gives the conclusion.

2 Related Work

Millions of users are sharing resources of cloud by submitting jobs to cloud servers. Consequently, scheduling millions of those tasks is a tough challenge to the cloud. The major area of researches lies in balancing the workloads among various resources and improving the QoS parameters like response time, total time of execution and throughput. To overcome this, several works have been proposed in the literature. This section describes static and dynamic load balancing algorithms, and the vivid analysis of honey bee's algorithm is emphasized along with its strength and weaknesses in load balancing. Also, this section demonstrates the effectiveness of self-aggregation (SA) in active clustering algorithm to improve the performance of load balancer.

2.1 Static and Dynamic LB Algorithms

Weighted round robin (WRR) [10] is a static LB and is designed to surpass the major issues faced by the RR algorithm. In weighted RR algorithm, each host has weight, and based on the magnitude of weights, tasks are assigned. The weights are accredited based on the resource availability. Therefore, the host with more resources has higher weight, thereby receiving more number of tasks. In Min–Min LB algorithm, tasks with lesser execution time are identified. From this, the algorithm selects the task that has the least execution time. For the remaining tasks, their execution time are updated for that host. This procedure is continued till every task is assigned the host with sufficient resource. However, the starvation problem, due to more number of tasks with higher execution time, affects the performance of this algorithm. In Max–Min LB algorithm [6], task which possesses maximum execution time is selected and allotted to the host. Better results are found with a static hybrid algorithm, after combining the weighted RR algorithm with the Max–Min LB to minimize the limitations (waiting/starvation) of both the algorithms. Equally spread current execution (ESCE) is a dynamic LB algorithm [6] and considers the task priority which is based on the span of task. It disperses load arbitrarily by checking task's length and then assigns the job to a host that has lesser load. Throttled is also a dynamic LB and assigns tasks in a first-fit basis, whereas modified throttled assigns tasks to the resources using the similar approach used by the next fit algorithm. In both the scenarios, the resources are not utilized properly.

Ant colony optimization (ACO) VM migration [9] overcomes the difficulties in migrating VMs via monitoring through local agents. It computes the utilization of VM of each host. After identifying the overloaded machines, they record them with prioritized sequence, and thereafter, it maps the sorted VMs. A modified ACO is proposed in [11] for distributing the workloads and resource properly in cloud environment. It is a random search approach for scheduling incoming tasks among VMs. It used a feedback mechanism. A head ant is defined to inform other ants about the status of hosts for load redistribution using forward and backward movement strategies. However, these algorithms fail to give good scalability and throughput [12] as the overhead is increased by more control parameters for mapping pheromone quantity and probability for each move.

2.2 Honey Bee Behavior-Based Algorithm

This section gives a detailed review on honey bee algorithms from the past studies which is based on load balancing and highlights the effectiveness and inadequacy of those algorithms. Cost-effective honey bee LB algorithm [13] aims for selecting the most optimal VM by estimating task's monitory cost and overall execution time in the selected VM. It chooses the one with the minimum time for execution and monitory cost. But, this LB algorithm causes a greater number of migrations that decrease the quality and the performance of overall system. Enhanced bee LB algorithm [14] transfers tasks from overutilized VMs to the most appropriate underutilized VMs. The approach fetches tasks with primarily less importance from the overutilized VM's queue and transfers to the underloaded VM. It guarantees that no tasks will be delayed for a longer time to get processed. However, task migrations are high, thereby affecting the overall performance. Honey bee behavior inspired load balancing [15] algorithm balances the workload among all VMs and minimizes the overall time of execution in cloud environment. Based on load, VMs are categorized as overutilized VMs, underutilized VMs and balanced VMs. Based on the load, tasks count and VM count in groups, the tasks are migrated from overutilized VM to underutilized VM. It has been shown that the average response time is enhanced, and waiting time on queue is reduced.

Enhanced honey bee aims to balance workloads in VMs while allocating task by examining the variations in individual VM processing time from the average processing time of all VMs. Once it exceeds a threshold (predefined value), the corresponding VM becomes overloaded at this time. Then, the algorithm starts the process for load balancing, which limits allocation to overloaded VM. Then, task will get executed to the completion without preempting and thus is referred as a non-preemptive system. The empirical results show that, for the HBA algorithm, increasing the server resources does not improve its equivalent throughput, as its performance remains consistent. On the contrary, the HBA progressively showed increase in performance when diversity in workload increases in accord with the server's workload. With these algorithms as a background, the proposed algorithm incorporates the self-aggregation to further enhance the performance as discussed in the next subsection.

2.3 Self-aggregation

This passage portrays the strategy of self-aggregation (SA), uncovering its positive and negative viewpoints based on the examinations in which the algorithm was validated. SA powerfully rewires the network. The algorithm focuses to group services which are of comparable perspectives together [16]. The greater part of the LB functions admirably in just a few circumstances when the nodes have knowledge of comparable nodes and delegate outstanding tasks at hand to them. The past investigations that demonstrated the standard rewiring process in this circumstance could certainly fix the issue, bringing about an ideal collaboration in the long term for similar procedures.

Evaluation results of certain experiments showed that the performance of SA increases as the number of hosts increases, whereas the performance of SA degrades with the diversity of hosts increase. To mitigate this particular issue, HB algorithm can be adopted. Most of the existing works did not try to combine some of these algorithms to root out their weaknesses. The main focus of these works is on changing the parameters of the different honey bee algorithms [17, 18]. All these studies reviewed so far did not combine the algorithms but did suggest combining them and improving the algorithm on their weaknesses. Within a limited size of resources, the increased size of system does not lead to increase in throughput performance; therefore, an improved HB algorithm is suggested to deal effectively with the issue. This review follows the research question where there is necessity to take up SA to improve the HB algorithm. A similar idea in [19] developed a comparable plan where SA and a blend of distinct algorithms have been implemented.

The conclusion from their research shows that clustering had an overhead effect; however, using a specialized algorithm besides SA has evolved out with good improvement in results. The functional analysis of SA algorithm is mentioned by highlighting its potency along with its shortcomings. Thus, the combination of algorithms is needed for collaboration to solve the primary issues of load balancing algorithms.

3 Proposed Enhanced Honey Bee (EHB) Load Balancing Algorithm

The proposed EHB is entirely inspired through the foraging nature of honey bees. The tasks that are assigned to VM accord information such as VM availability status and workload of VMs and the number of waiting tasks. This is similar to how the bees inform/update other bees in the beehive with their waggle dance on finding an abundant source of honey. The objective in the proposed EHB model is to distribute workload by avoiding underutilization and overutilization of resources. It allocates incoming tasks to VM on meeting the two criteria: (i) Task count of currently processing VM should be less than task count of all other VMs. (ii) The deviation of

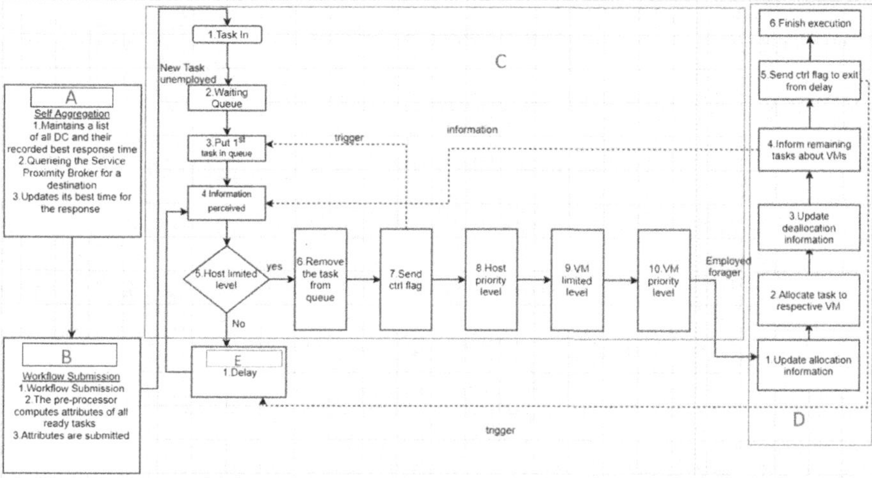

Fig. 1 Block diagram of the proposed EHB

processing time in the current VM from the average processing time in all VMs should not exceed a threshold value. The detailed behavioral control design structure of the proposed EHB is shown in the Fig. 1.

The proposed EHB algorithm consists of five stages called A, B, C, D and E. The five stages of the EHB algorithm are discussed in the following paragraphs.

Stage A: Self-aggregation (SA): SA technique is a load balancing technique that groups services together depending on certain characteristics. The SA algorithm is implemented on a Service Broker component of cloud analyst. At this level, the load is re-configured dynamically by monitoring the best optimal response time achieved for that user, proximity of UB and the current workloads of data centers (DC). The major objective of this algorithm in the broker policy is to have level control over user request sending to DC. This is mainly to address the issue raised in the background section where the honey bee's performance degrades as requests count and system size are increased. Its flow diagram is depicted in Fig. 2.

Algorithm for Self-Aggregation (SA):

1. A list with the record for every DCs is maintained, and also, a list contains all the recent/current best time of requests response for each individual DCs.
2. Once the Internet receives any message from userbase, it consults self-aggregation algorithm to determine the best appropriate data center for the request.
3. SA algorithm then queries the service proximity broker for a destination.
4. SA algorithm then updates its best time for the response if the new response time is better than the previous.

Stage B: Workflow Submission: (The input is a specified group of tasks) A group of tasks called workflow is given as input. This stage assembles tasks to move individually through subsequent stages after computing values for workflow char-

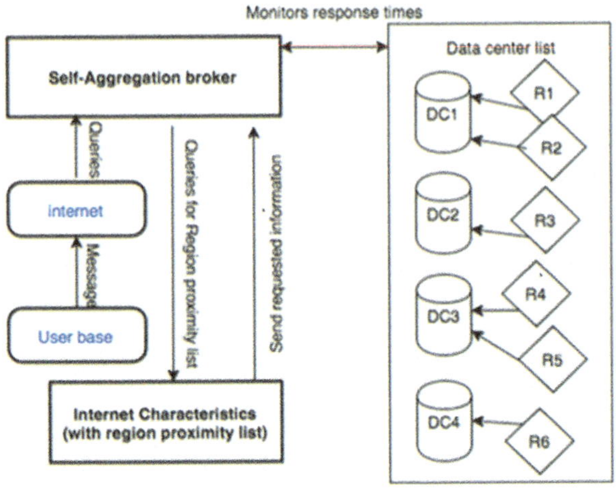

Fig. 2 Block diagram for self-aggregation

acteristics. The tasks count and instruction length for individual task are computed. These values are used for two chief functions, one to prioritize among the VMs and second to avoid the overutilization of VMs. The allocation strategy focuses on the priority information content, i.e., load of each VM which confers the tasks length. Model limits more allocation to overloaded Vms which have higher processing time than a threshold in terms of variation value. The steps involved in this stage are given as follows:

1. If the workflow is arrived, then it is directly submitted to the preprocessor.
2. In the next step, the workflow characteristics are computed by the preprocessor, i.e., the tasks total count and its instruction length.
3. After that, the task attributes information is stamped to the task.

Stage C: Task in: (The input is an individual task) This stage is the heart of the proposed EHB algorithm, and its steps are elaborated as follows:

1. All tasks, which are ready for execution, send a request signal to VM LB.
2. The VM LB puts the ready tasks into the waiting queue.
3. Fetching order of tasks is in first-come-first-serve (FCFS) order. Then, the VM LB fetches the first task from the waiting queue.
4. Determine the suitable VM for which the task needs to be assigned by collecting information subsequently after the last allocation and de-allocation. The procedure is exactly same as to which flower patch honey bee should visit depending on honey is present in the flower or not. Information perceived is of two types.
   (a) Threshold information which measures availability for the hosts depending on variation of the hosts. The same availability check is implemented in VM level.

(b) Priority information contains information about the current workload. Load indicator is taken as processing time in host level and is considered as the tasks count in VM level.

5. Limit level (host): It limits the request allocation to the overutilized hosts. The task fetched from waiting queue has to find available host for the allocation. It has two possibilities based on threshold information, either it finds record of available hosts or then checks priority information to choose the host with the least load. If it is not present, then it will be delayed until host becomes available.
6. The first task is selected from waiting queue if the task finds available hosts.
7. Control flag is sent to get another task from waiting queue.
8. Priority level (host): There is a possibility of having more than one hosts that may be ready to accept the task. If so, the task chooses appropriate host based on priority information. It should consider the available host with minimum load. In this level, the task is mapped to specified host based on priority information of hosts.
9. VM limited level: Finding available host means that there should be atleast one available VM. Then, the rule according to this level determines the record for VMs available in a specified host.
10. VM priority level: There may be more than one available VMs in the specified host which can accept the specified task. Thus, the task is choosing desired VM depending on priority information. The priority information in this level is the tasks count handled by VMs.

Stage D: Task out: (After choosing suitable VM)

1. The information about the allocations is updated by the tasks, i.e., count of tasks that the VM is processing, tasks processing time and monitors loading status of host and VM.
2. VM(j) becomes overloaded depending on the variation/dispersion of processing time, i.e., this variation value gives direct indication of load balance between this VM(j) and other VMs. This is due to the variation in the current VM's processing time from meantime of processing in all VMs that exceeds a threshold value. When every VMs in a specified host become overloaded, this host is treated as overloaded.
3. The task updates de-allocated information. State of VM(j) becomes available when variation in the currently available VM's time of processing the request from average time of processing in all VMs does not exceed threshold valuc.
4. Informs the remaining tasks in queue regarding status (host, VM) in a way the bees update other bees with waggle dance about the source of honey. The information update assists directly the LB in deciding which VM to assign to the current request/task.
5. Sends control flag to tasks to exist from the delay unit when allocated task has finished execution.
6. The task finishes execution.

Stage E: Delay: If task does not find any available host, it goes for delay stage until it receives a control flag from task which has finished its execution. Then, the task starts from earlier point, and it has checked the information perceived. The delayed task is served first before the tasks waiting in the queue. Each task after the first round has to go to Step B.4 and then through all the remaining steps. It benefits from the knowledge collected by "bees" for selecting more appropriate host and subsequently suitable VM for allocation. In addition, situations where all hosts become overloaded, the tasks have to enter a delay until they receive information from "bees" about any host becomes available.

4 Simulation and Performance Analysis

The proposed EHB model is simulated using cloud analyst tool in large-scale cloud environment, and the results are analyzed and compared with three conventional static and dynamic algorithms and two SI algorithms: RR, ESCEL, throttled, ACO and PSO, respectively. This section deeply examines the analysis methodologies and tools that are used to analyze the collected data generated by the proposed solution.

4.1 Simulation Setup

To evaluate efficiency of EHB designed mode, the virtual nodes and computing resources are modeled. The experiments are implemented with 10 data centers, 2–6 hosts, 50 VMs and 100–1000 tasks under the simulation platform. The task's length varies from 1000 to 20000 MI. Processing speed, available memory space and bandwidth determine the allowable load of each VM. The other parameters [15] used for the simulator are depicted in Table 1.

4.2 Simulation and Analysis

This subsection discusses the performance of the proposed EHB algorithm in terms of average response time and degree of imbalance. More specifically, the average response time of the proposed EHB algorithm is analyzed with (i) varying number of tasks, (ii) executable instruction length (IL) per task, (iii) conventional algorithms like RR, ESCEL and throttled algorithms and (iv) SI-based scheduler algorithm.

4.2.1 Average Response Time Versus Varying Number of Tasks

The average response time expresses the average time spent between request submission and getting the first response for request. Figure 3 shows the average response

Table 1 Simulation parameters

Group	Parameters	Range of values
Task (cloudlet)	Task's length (executable instruction length(bytes))	1000–2000
	Total count of tasks	100–1000
VM	Total VMs used	50
	Processor speed	500–2000 MIPS
	Available memory space in a single VM	256–2048 Mb
	Bandwidth	500–1000
	Scheduler used	Time shared
	Count of the processor elements (PEs) requirement	1–4
Data center	Total datacenters used	10
	Number of hosts	2–6
	Scheduler (VM)	Time shared

Fig. 3 Average response time versus number of tasks

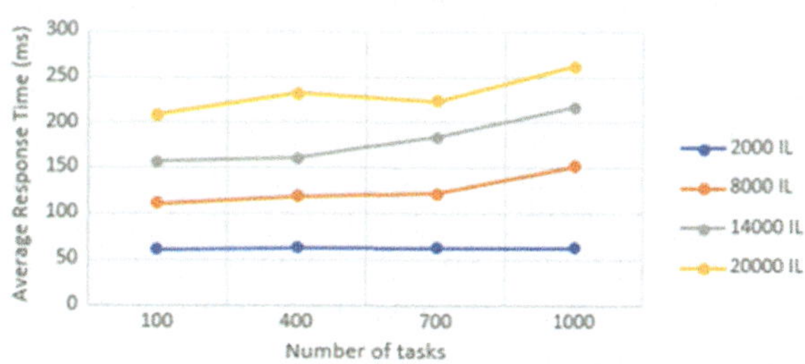

taken by the proposed EHB for varying number of tasks with different executable instruction lengths (ILs).

It is observed from this figure that the average response time increases slightly at low average count of requests; however, it becomes very fast at high average number of requests. However, the increase in instruction length greatly affects the response time as it increases system's load. Therefore, the factors such as availability and load of individual VM are utmost crucial for task allocation.

4.2.2 Average Response Time Versus Executable IL Per Task

Figure 4 shows graph of average response time versus IL at varying values of requests count. Initially, it starts between 62.19 and 263.50 ms at all situations but increases tremendously when requests count and executable IL are increased simultaneously. This is as because the system's load is increased heavily which effects the overall time of response.

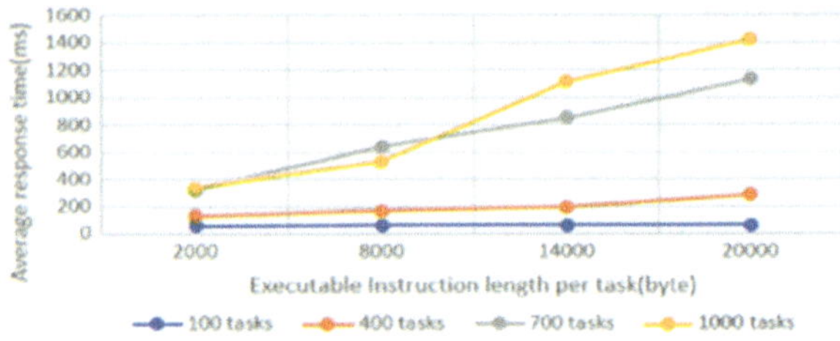

Fig. 4 Average response time versus number of tasks at varying IL

4.2.3 Comparisons with Conventional Algorithms

From the previous results, it is clear that the executable IL and the requests count influence the overall system performance. Simulation results for the proposed EHB model are compared with conventional static and dynamic algorithms such as RR, ESCEL and throttled algorithms with varying number of tasks. Instruction length is arbitrary changing values between 100 and 1000 that lead to a variation of load among different VMs in cloud system.

Figure 5 shows the average response time achieved by RR, throttled, ESCEL and EHB algorithms versus tasks count. The tasks instruction length is arbitrary changing from 2000 to 20000. It is shown that EHB is saving up to 50% of average time of response over the other algorithms. The meantime for response of the EHB reaches about 1090 ms when the tasks count is 1000 and 2098 ms at other algorithms. This is because the EHB considers least load, availability of VMs and load variation of individual VM while allocating VMs to tasks. The graph also depicts that at all values, the meantime for response is between 360.29 and 462.19 ms but increases tremendously as the tasks count increases.

4.2.4 Comparison with SI Algorithms

Figure 6 depicts the average response time of the proposed EHB, PSO and ACO algorithms. It is inferred from this graph that all the values of average response time lie between 53.26 and 90.08 ms but increases tremendously as the tasks count

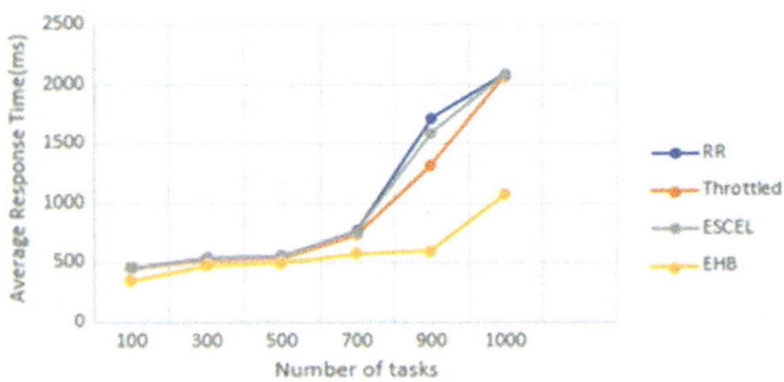

Fig. 5 Average response time versus number of tasks

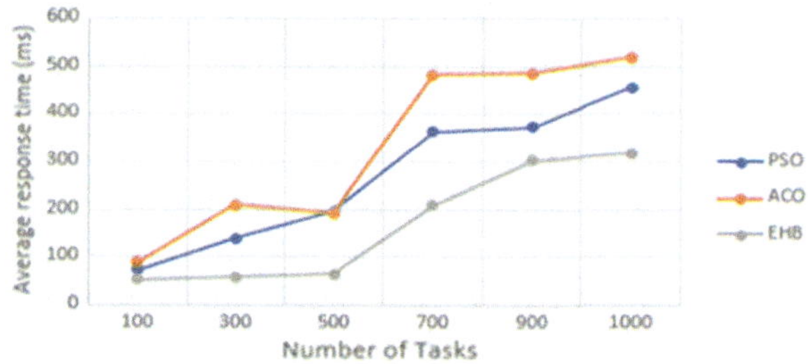

Fig. 6 Average response time versus number of tasks

increases. Since the proposed EHB algorithm allocates the tasks by considering the current status of VM and load variation, the average response time is minimized compared to that of PSO and ACO algorithms.

4.2.5 Imbalance Degree

DI measures imbalance among the VMs, whose value is evaluated through the equation as follows:

$$DI = \frac{PT_{max_{vm}} - PT_{min_{vm}}}{PT_{avg_{vm}}}$$

The degree of imbalance attained by the proposed EHB algorithm, PSO, ACO, RR, ESCEL and throttled is depicted in Fig. 7. The proposed EHB aims to minimize DI. While allocating VMs, considering this metric helps to preserve the balance of workloads among VMs. It is clear from the graph that the proposed EHB highly preserves the DI since EHB limits allocation of requests to VM when variation in VM's time for processing requests exceeds value of threshold. This assures to give the least

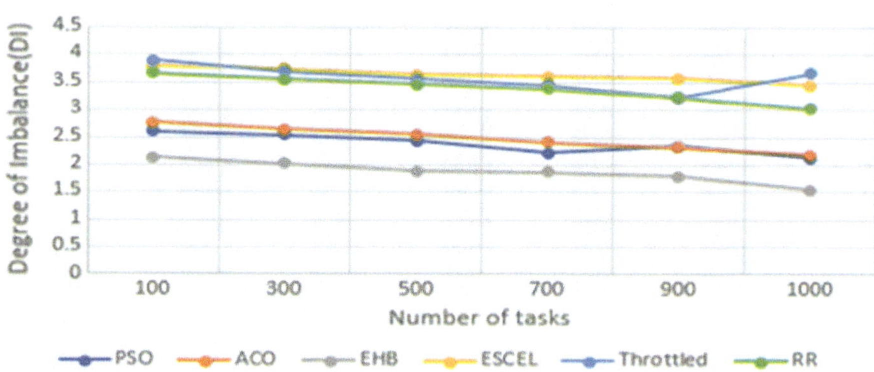

Fig. 7 Degree of imbalance versus number of tasks

value of DI compared to other algorithms because these algorithm makes the balance when the overall system becomes unbalanced which increases the imbalance degree.

4.2.6 Comparisons of Standard Deviation

Figure 8 shows the bar graph for load SD which conveys load balancing among VMs. Value of threshold is changed from 0.0 to 1.0; it shows exact value of threshold that correspondingly gives the least deviation as 0.1. It is because threshold value represents standard normal deviate for individual VM's time for processing request, and therefore, decreasing threshold value leads to decreasing SD of VM load. However, when threshold value is zero, the load deviation is increased since it signifies that

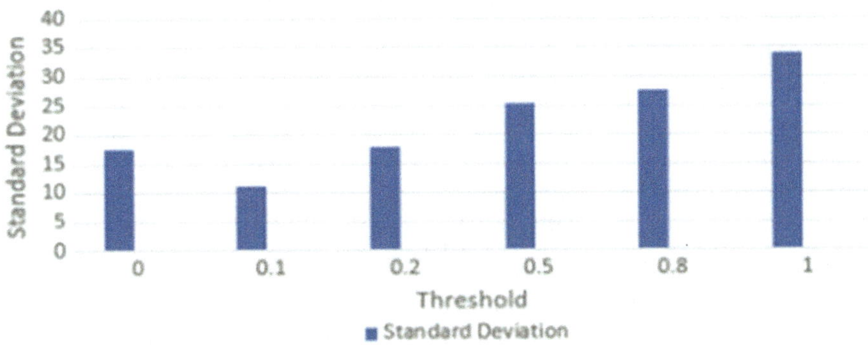

Fig. 8 Standard deviation of EHB with different values of threshold

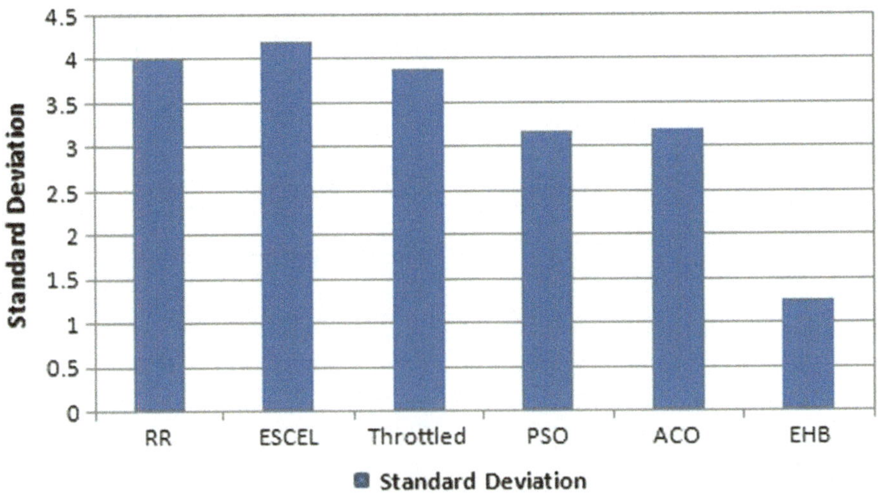

Fig. 9 Comparison of SD between different algorithms

effect of threshold condition is inactive in this case. Therefore, the value 0.1 is taken under consideration when simulating the proposed algorithm.

Figure 9 depicts the graph showing the comparison of standard deviation of EHB, PSO, ACO, RR, ESCEL and throttled algorithms with the different tasks count. It is clear from the figure that the SD [15] of the proposed EHB gives enhanced result when compared with other algorithms since it adapts a value of threshold that preserves and gives minimal deviation.

5 Conclusion

Enhanced honey bee (EHB) algorithm is proposed to optimize resource allocation and load balancing in a vast fluctuating cloud scenario. This proposed algorithm follows the foraging nature of honeybee as it gets a similar resemblance to resource allocation with balancing load in heterogeneous cloud. The proposed EHB aims to decrease overall response and data center time to process requests by distributing workload among different VMs. It limits allocation of requests to VM when the variation in VM's time exceeds a predefined threshold. The EHB algorithm has shown better simulation results which improves the response and data processing time over the well-known algorithms such as RR, throttled and ESCEL. In addition, it preserves deviation and balancing effectively than the existing SI algorithms, ACO and PSO.

In the future, the proposed EHB algorithm can be analyzed in big data applications [20]. Even though tasks are clustered based on the best time for response to DC using self-aggregation, there is need for tasks in group to be served inside VM LB. Such that, all tasks can surpass the constraints at one time and optimize the time and deviation even more as it will eventually reduce the computational time.

References

1. Randles M, Lamb D, Taleb-Bendiab A (2010) A comparative study into distributed load balancing algorithms for cloud computing. In: 2010 IEEE 24th international conference on advanced information networking and applications workshops (WAINA), pp 551–556. https://doi.org/10.1109/WAINA.2010.85
2. Saffre F, Tateson R, Halloy J, Shackleton M, Deneubourg JL (2009) Aggregation dynamics in overlay networks and their implications for self-organized distributed applications. Comput J 52(4):397–412. ISSN 0010-4620. https://doi.org/10.1093/comjnl/bxn017
3. Sahu Y, Pateriya RK, Gupta RK (2013) Cloud server optimization with load balancing and green computing techniques using dynamic compare and balance algorithm. In: 2013 5th international conference on computational intelligence and communication networks (CICN), pp 527–531. https://doi.org/10.1109/CICN.2013.114
4. Topcuoglu SH, Wu M-Y (2002) Performance-effective and low-complexity task schedulingfor0 heterogeneous computing. In: IEEE transactions on parallel and distributed systems, 13(3):260–274. ISSN 1045-9219. https://doi.org/10.1109/71.993206

5. Vasile M-A, Pop F, Tutueanu R-I, Cristea V, Koodziej J (2014) Resource-aware hybrid scheduling algorithm in heterogeneous distributed computing. Futur Gener Comput Syst 51(0):61–71. ISSN 0167-739X. https://doi.org/10.1016/j.future.2014.11.019
6. Rastogi G, Sushil R (2015) Analytical literature survey on existing load balancing schemes in cloud computing. In: International conference on green computing and internet of things (ICGCIOT), pp 1506–1510
7. Sesum-Cavic V, Kuhn E (2010) Comparing configurable parameters of swarm intelligence algorithms for dynamic load balancing. In: 2010 Fourth IEEE international conference on self-adaptive and self-organizing systems workshop (SASOW), pp 42–49. https://doi.org/10.1109/SASOW.2010.12
8. Abu-Mouti FS, El-Hawary ME (2012) Overview of artificial bee colony (abc) algorithm and its applications. In: 2012 IEEE international systems conference (SysCon), pp 1–6. https://doi.org/10.1109/SysCon.2012.6189539
9. Dorigo M (1992) Optimization learning and natural algorithms. Politecnico di Milano, Milano Ph.D. Thesis
10. Panwar R, Mallick B (2015) Load balancing in cloud computing using dynamic load management algorithm. In: Proceedings of the 2015 international conference on green computing and internet of things (ICGCIoT), pp 773–778
11. Wen W-T, Wang C-D, Wu D-S, Xie Y-Y (2015) An ACO-based scheduling strategy on load balancing in cloud computing environment. In: Ninth international conference on frontier of computer science and technology, pp 364–369
12. Sharma KNP, Krishna V, Gupta C, Singh KP, Nitin N, Rastogi R (2012) Load balancing of nodes in cloud using ant colony optimization. In: Proceedings of the 14th international conference on modelling and simulation, pp 3–8
13. Sheeja YS, Jayalekshmi S (2014) Cost effective load balancing based on honey bee behavior in cloud environment. In: Proceeding of first international conference on computational systems and communications (ICCSC), Trivandrum, pp 214–219
14. Babu KR, Samuel P (2015) Enhanced bee colony algorithm for efficient load balancing and scheduling in cloud. In: Innovations in bio-inspired computing and applications, advances in intelligent systems and computing, vol 424. Springer, pp 67–78
15. Hashem W, Nashaat H, Rizk R (2017) Honey bee based load balancing in cloud computing. KSII Trans Internet Inf Syst. 11(12):5694–5711. https://doi.org/10.3837/tiis.2017.12.001
16. Di Nitto E, Dubois DJ, Mirandola R (2007) Self-aggregation algorithms for autonomic systems. In: Bio-inspired models of network, information and computing systems, pp 120–128. https://doi.org/10.1109/BIMNICS.2007.4610096
17. Randles M, Lamb D, Taleb-Bendiab A (2009) Experiments with honeybee foraging inspired load balancing. In: Second international conference on developments in esystems engineering (DESE), pp 240–247. https://doi.org/10.1109/DeSE.2009.19
18. Randles M, Odat E, Lamb D, Abu-Rahmeh O, Taleb-Bendiab A (2009) A comparative experiment in distributed load balancing. In: 2nd international conference on developments in esystems engineering (DESE), pp 258–265. https://doi.org/10.1109/DeSE.2009.20
19. Xu M, Tian W, Buyya R (2017) A survey on load balancing algorithms for virtual machines placement in cloud computing. Concurr Comput: Pract-Exp 29(12):e4123
20. Aceto G et al (2019) Know your big data trade-offs when classifying encrypted mobile traffic with deep learning. In: IEEE network traffic measurement and analysis conference (TMA)

Real-Time Communication Alert System for Missing Vessels in Deep Sea

M. T. Chitra, B. Gayatri Menon and Elizabeth Sherly

Abstract Understanding the paradox connected with the movement of marine objects has always been a key issue in many areas of communication and other scientific investigations. In developing countries, capsize of boats and missing vessels still top up daily news headlines. Competence of data mining techniques has been widely utilized in various applications to uncover hidden patterns, knowledge extraction and prediction effectively. However, none of such systems are found for analysis of marine data to track and manage information of fisherman and fishing fleets, which has to be attended seriously for effective communication alert dissemination and also to secure them during emergency situations. The work proposes a real-time communication alert system for finding missing vessels in deep sea. The potential of machine learning-based approaches is utilized in this project; fuzzy c-means clustering is used for localizing the vessels in deep sea. Information theory-based parameter selection is used for finding the parameter that influences the movement of the missing vessel and predicting the location to which missing vessel has drifted toward. Knowledge of past trajectories of the vessel is also considered for better spotting of the missing vessel. A decision tree-based classification model is applied along with the proposed clustering for finding the best vessel nodes through which the communication alerts can be extended to save and secure the life of fishermen in time. A performance comparison study of the decision tree and artificial neural network-based classification for identifying best vessel for communication is also done.

Keywords Machine learning · Clustering · Missing vessels · Information theory · Fuzzy c-means · Decision tree classification · Artificial neural network

M. T. Chitra · B. Gayatri Menon (✉) · E. Sherly
Indian Institute of Information Technology and Management-Kerala, Technopark, Trivandrum, India
e-mail: gayatri.menon@iiitmk.ac.in

M. T. Chitra
e-mail: chitra@iiitmk.ac.in

E. Sherly
e-mail: sherly@iiitmk.ac.in

S. M. Thampi et al. (eds.), *Applied Soft Computing and Communication Networks*, Lecture Notes in Networks and Systems 125,
https://doi.org/10.1007/978-981-15-3852-0_13

1 Introduction

The resource abundance and position of fishing fleets are increasing in significant numbers. At the same time, the names on a list of lost or capsized vessels grow by hundreds in each year around the world. When an emergency or accident occurs, the search and rescue system (SAR) associated within the country swiftly becomes the primary safety net. The emergence of location-based positioning devices aids in tracking and recording the moving vessel trajectories more easily [1]. However, the uncertainty in sea environmental conditions prospects poor network connectivity issues with these devices, which can adversely affect the trajectory tracking process of the fishing vessels. In addition, reception of good mobile communication signal ranges in deep sea is diminutive, especially during rough weather conditions, and hence, locating the isolated fishing vessels when there is no communication possible with the SAR systems becomes more strenuous. This unfolds the necessity of inventing intelligent systems which help in tackling such emergencies by identifying vessel clusters to which communication can be extended and thereby saving the precious life in time. The strengthening of the existing communication technologies is one of the major prerequisites to widen the exploitation of deep-sea fisheries.

The existing works available on literature on account of trajectory analysis and location prediction of the moving vessels in maritime rely on either automatic identification system (AIS) or vessel monitoring system (VMS) [2]. The complexity and large quantity of AIS or VMS data will create difficulties for the commercial software systems for locating the missing vessel trajectory effectively in real time [3, 4]. However, the unavailability of accessibility of such systems for fishing vessels in Indian coastlines due to numerous security reasons is the real motivation toward this work. Predicting the missing fishing vessel locations in the deep sea is a challenging task in maritime surveillance systems as it demands the influence of the mobility parameters of the fishing vessels, the trajectory followed by the vessel, vessel speed, direction and their dynamics over time accompanying several environmental parameters of the sea such as wind speed and direction, current direction, tidal height, sea surface temperature and so on [5]. The mobile communication parameters like signal strength, transmission rate, reception rate of the mobile or the wireless devices also play a distinct role in the proper identification of these vessels at the time of its missing. Incorporating these characteristics, when available, can help predict the missing vessel location, but adds complexity to the tracking process. A real-time communication alert system for missing vessels in deep sea is the need of the hour. Data mining approaches can adequately analyze the location-based data of the fishing vessels, thereby producing sustainable communication and intelligent information dissemination systems. It reinforces the decision support systems with sensible knowledge in discovering user context, real-time mobility prediction of mobile nodes, especially during emergencies like disaster alerts, missing vessel identification, etc.

The work presents an intelligent decision-making system for handling emergencies of fishing vessels in deep sea. The work explicitly focuses on tracking and locating the trajectory data of the fishing vessels using the GPS data and mobile network connectivity using proposed machine learning algorithms [6, 7]. The vessel locations are traced using the GPS connected to these moving vessels, which helps in generating the trajectories of these vessels while they are at sea. The past trajectory patterns of the particular vessel are also taken into account for reliable prediction [8]. The trajectory patterns or behavioral models of the individual fishing vessels and the bounding trajectory can represent a generalized set of movement patterns that the vessel may use to follow in deep sea during its entire trips, which help to locate its current position in the sea. It facilitates storing and analyzing the pattern of movements of individual fishing vessels, thereby finding any kind of invasion for ensuring maximum security of the life of the fishermen. This tracking facility is economical and affordable for the fishermen community as it uses local mobile networks for communication and alerts. Intuitively, it initiates the rescue operations by predicting suspected regions where the missing vessel is expected to be found by applying hierarchical clustering [9] of the vessels available in the region, thereby identifying the vessel clusters in those specific locations and send them alerts regarding the missing vessel. The spatiotemporal-based approach followed in this work helps in achieving a more accurate prediction of the suspected locations. The experimental results showed the effectiveness of the proposed system on vessel data collected from various coastal ports in Kerala, India.

The paper is organized as follows: Sect. 2 contributes a brief review of literature in the domain of mobility tracking algorithms and identification technologies. Section 3 gives a detailed description of the proposed methodology and architecture. Section 4 discusses results and findings. Section 5 includes the concluding remarks of the work.

2 Related Works

Advancements in the location tracking and identification technologies help in collecting the trajectory data of moving objects in an efficient manner which paves way for finding the essential information from the trajectory dataset [10]. A lot of research work is available in the existing literature on trajectory data analysis of terrestrial environment [11–19], which proposes several interesting algorithms for mining the movement patterns from these trajectories. However, to obtain similar movement patterns from the trajectory dataset of marine data trajectories, assort the similar trajectories as clusters. A conventional clustering approach could be of greater use to depict the trajectories with matching feature attributes. At the same time, it is too challenging to define a uniform clustering criterion of these trajectories owing to the complex and the unforeseeable nature of the sea attributes.

Automatic Identification System (AIS), which uses satellite communication channel, attains a global reception which discovers the detailed information with respect to each vessel, including its unique ID number, speed, course over ground and its direction of travel to the coastal maritime authorities as well as other ships on a timely basis thereby making it possible in tracking and monitoring the ship's movement and behavior effectively [3]. However, this is very expensive and cannot be afforded by the fishermen community in India.

This research work, hence, utilizes the efficacy of clustering algorithms to analyze the mobility patterns of vessels in the sea, especially in a present Indian scenario where AIS/VMS systems have license restriction and expensive. Clustering algorithms are commonly categorized as partition-based methods such as K-means, K-Medoids, CLARANS algorithms [20], hierarchical clustering methods such as BIRCH algorithm [21], density-based clustering methods such as the DBSCAN [22], OPTICS [23] algorithms and STING algorithm which is a grid-based clustering method [24].

Gaffney and Smyth [25] proposed an algorithm for classifying the trajectories with comparable features by applying regression mixture model and the expectation–maximization algorithm. The trajectory clustering is performed by estimating the total distance between two complete trajectories. However, missing of some general characteristics in sub-trajectories while considering the entire trajectory set was one of the major difficulties faced in this work. Lee et al. [26] in their work suggested a framework which can partition the whole trajectory into several segments, and arrangement of nearby trajectory segments into clusters is done. The Hausdorff distance measure is used for clustering the trajectories which find a representative path for each cluster of segments. Li et al. [27] suggested an algorithm which processes one data element at a time in an incremental fashion, which can help to reduce the cost of computation and repository of received trajectory data.

Knorr et al. [28] adapted similarity-based measures to estimate the attributes of vessel information such as source and destination locations, direction of travel and speed in the maritime domain. The distance-based similarity measure between two trajectories was calculated using the summated result obtained for the feature sets. Bomberger et al. [29] resolved the whole region of interest area into distinct grid locations and specified a grid-based method to investigate the vessel movement behavioral patterns for maritime monitoring. Dahlbom and Niklasson [30] mentioned a density-based clustering method for identifying vessel behavior in coastal surveillance systems. The detected vessel trajectories are processed to generate knowledge which represents the normal path of vessel traversal. Auslander et al. [31] explained several types of maritime traffic with respect to different levels of complexity and suggested some local anomaly detection algorithms, whose processing may vary depending on the type of maritime traffic.

Vespe et al. [32] in their work developed a waypoint density algorithm that can reduce surplus volumes of AIS dataset as a list of appropriate route points and outline waterways that subsequently traverse through a series of lines connecting waypoints. Pallotta et al. [33] developed a density-based algorithm that can group the vessel trajectories as route, waypoint and stationary objects. The work suggested a framework which can discover the abnormal vessel behavior and can also predict the motion of the ship.

Liu et al. [34] suggested a method that employs speed and direction information to identify move and stop location points of the vessels, thereby segregating the cluster head of the clustered region. Lei [35] modeled the sailing patterns of vessels taking into account the static sequential, temporal and spatial aspects of the vessels. An aggregated index of outlying scores to identify abnormalities in vessel behavior has also been defined here. In order to detect the frequent regions, a grid-based clustering approach is used which defines feature points to represent the cell size and decrease the loss problem.

Zhen et al. [36] suggested an approach for detecting maritime abnormalities using a mixed approach. Clustering of vessel trajectories by introducing a structure similarity measurement for determining space and directional similarity and an optimized combination parameters using naïve Bayes approach was also measured.

Most of the systems described above are for monitoring merchant vessels and ships and uses the satellite-based communication for tracking and operation monitoring. Here in our work, we are utilizing the potential Global Positioning System (GPS) and mobile communication for tracking small fishing boats and vessels.

3 Proposed Methodology

Transforming the mobility data into knowledge patterns that formulate a collective movement behavior, pertaining to groups of individuals, is worth putting into evidence. The work aims at predicting the expected locations where the missing vessels can be found by understanding its general movement behavioral pattern along with environmental conditions. The location information and trajectory information of the moving fishing vessels serves as the knowledge base which is being exploited in the proposed location prediction algorithm.

The knowledge discovery process presented in this paper helps in tracking and predicting the position of the missing vessels by analyzing the live vessel movement data collected from a GPS-enabled Android mobile phone used in the vessel along with the historical mobility pattern of the specific vessel. The limited communication range of the mobile device, disconnection, speed and direction of travel of the vessel, location error caused by environmental parameters of the seabed such as wind speed, tidal height also influences the prediction process.

The overall workflow of the proposed approach is shown in Fig. 1.

The previous trajectories of the vessels in the sea also play a crucial role in analyzing the mobility pattern of a vessel moving in a specific travel path. The mobility pattern constitutes the general behavior of a group or subgroup of trajectories or paths of the moving fishing vessels recorded at regular time intervals with location (space) and time information, described as a tuple of $<(x_1, y_1, t_1), (x_2, y_2, t_2), \ldots, (x_n, y_n, t_n)>$; $t_1 < t_2 \ldots < t_n$, where x and y represent the geographical coordinates, longitude and latitude, respectively, in a specific time interval t_n. The trajectory basically denotes the entire route that the particular vessel follows in a specific period of time, say a day. Each point in the trajectory represents a location position in space at a certain instance of time. It is assumed that during each period $[t_i, t_{i+1}]$, the vessel at position (x_i, y_i) at time t_i moves along a shortest straight line from $(x_i,$

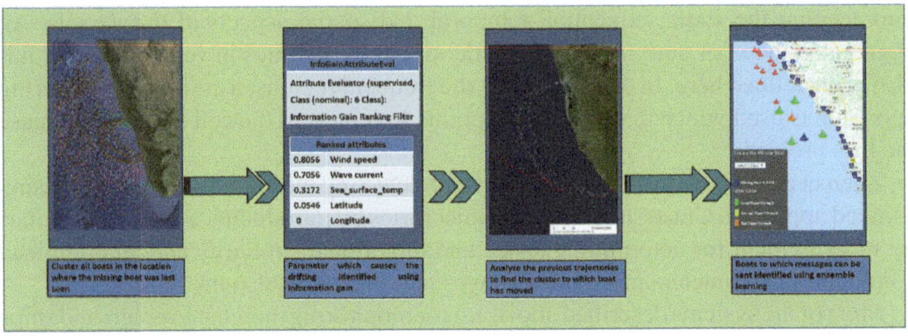

Fig. 1 Overall flow diagram of the proposed system

y_i) to (x_{i+1}, y_{i+1}) at a constant speed, and intermediate point is obtained by linear interpolation [37].

The observation of the movement patterns based on their direction of travel and the behavioral models of vessel helps comprehend the vessel mobility and sustainability, which in turn support in location prediction with trustable knowledge [37]. The extracted knowledge supplemented by data mining and machine learning algorithms helps in precisely identifying the location to which the missing vessel might have drifted [22].

Identification of the most probable locations where the missing vessel is expected to be found and locating the cluster heads available in those areas of interest are of prime importance in this work. All the vessels present in the suspected regions of missing are clustered using fuzzy c-means algorithm. Once the vessels are clustered, information gain-based attribute selection method is employed to discover the main parameters which are critically causing the change in the direction of normal movement of the missing vessel. Analysis of the past trajectory of the vessel in this particular route is also performed along with this step to find the direction to which the vessel has moved. Once the cluster in the suspected regions to which the boat has drifted is identified, a classification based on decision tree approach is applied to determine the best node (cluster head) among them to which communication can be extended to convey emergency message alerts to rescue the missing vessel in time.

3.1 Vessel Cluster Identification for Mobility Prediction

The variation in the Course Over Ground (COG) of each vessel is continuously monitored for finding the unusual scenario easily. Whenever a vessel missing case is encountered, the rescue services will get activated based on the information collected such as the last location of the vessel, the direction of its travel at the time of its missing and so on. The proposed mobility prediction algorithm initiates in identifying all the vessel clusters that are currently available in the expected locations where the missing vessel is supposed to be found and will intimate the identified head vessel clusters about the missing vessel.

The environmental conditions as well as the communication parameters of that location are considered to find the best vessel nodes in the cluster to which communication can be extended from the base station. The velocity of the missing vessel and the direction of its travel are also considered as a major factor while determining the expected trajectory of the vessels. The environmental parameters of the sea bed have a main role in determining the location of the missing vessel. The best suitable vessels in the communication pattern in which the affected vessel may appear will be taken and analyzed to obtain its mobility pattern in the sea. This in turn would help the emergency rescue team to localize the missing vessel at times of calamities. The machine learning approaches applied in the algorithm help in analyzing and classifying the communication parameters easily. This in turn helps in identifying the mobility pattern and thereby inferring the anticipated location of the missing vessel (Fig. 2).

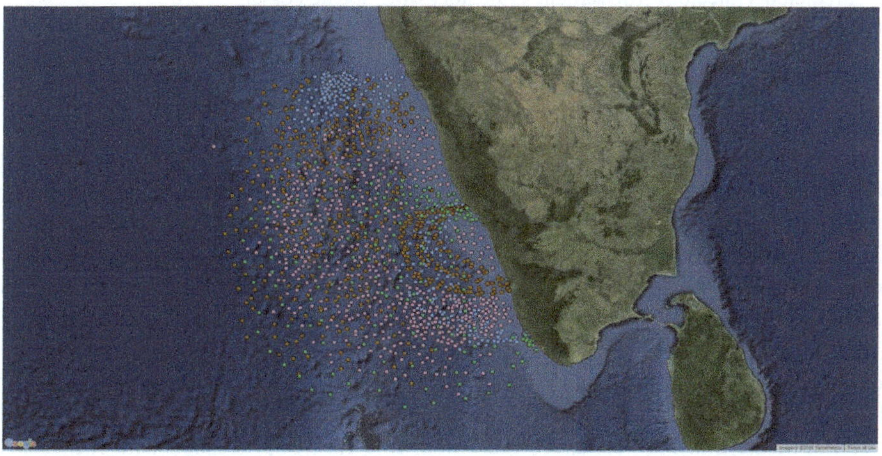

Fig. 2 Cluster of all vessels in the sea at a particular time stamp

Algorithm 1 Algorithm for location prediction of missing vessels based on its mobility trajectory.

```
BEGIN
locateVesselClusters(FishingVessels[], time){
Initialize Cluster := null;
update Cluster { TC_1, TC_2 ,TC_3 , ..., TC_n};
locateVessel(LAST_LOC (lat, long: time), env_par, vessel_speed, vessel_dir);
Update the vessel_cluster_db;
   apply FCM();
Invoke InfoGain_Attribute Selection(Lat, Long, Speed of the vessel, Direction, SST,
Pressure, Wind Speed, Sea surface current, Tidal height);
   Select Attribute(Info(D))==max
   Create a root node N
   Label node N with splitting_criterion;
   Assign the input trajectory path in to the appropriate cluster
C_k if there exists more than one such clusters
Choose the one with minimum distance subject to the constraint, Min (dist)
= Lower_bound <= (dist (TBi, C(TBj)) <= Upper_bound
   Predicted Trajectory=Movement Trajectory of C_k locateVessel
(LAST_LOC (lat, long: time), env_par, vessel_speed, vessel_dir)
   Find the probable bounding_region_set (lat,lon, radius)}
   END
```

Algorithm 1 explains the main procedures taken place during the localization of the vessel clusters in deep sea. The fishing vessel dataset is the main parameter supplied along with the main function *locateVesselClusters*(FishingVessels[], time). The *locateVessel()* identifies the best vessels locating near to the expected locations based on the input parameters supplied to it. From the last location, the vessel was seen, the direction of the travel and the speed of the vessel as major constraints, fuzzy c-means (FCM) clustering is initiated for identifying the nearby vessel clusters in the region of interest, which is the main activity of *apply*FCM(). Based on the Euclidean distance measure, those optimal clusters which are at a minimum distance from the location of the missing vessel are identified. It is further processed with the vessel parameters like the velocity of the vessel, moving direction at the time of its missing and also the sea bed data. *InfoGain_Attribute Selection*() method is then invoked for finding the high priority parameter which contributes to the movement of the vessel and find out the cluster and possible locations to which the vessel has drifted. The prediction estimation would help the SAR teams to localize the missing vessel at any emergency situation without direct contact.

3.2 Best Vessel Identification for Communication

In this work, C4.5 decision tree-based classification algorithm is utilized for locating best vessels in the identified cluster set. The potential of information gain is employed for selecting the most appropriate attribute that can well partition the dataset according to data collected from the fishing vessels. The input attributes list consists of latitudinal and longitudinal position of the boat, surface temperature of the sea bed, wind speed, tidal height, speed of the boat, direction of travel, signal strength of the mobile device, transmission and reception rate of the communication signal. The decision attributes in this context are *good*, *average*, and *bad* based on the overall communication capability and environmental condition of the vessel. To select the splitting attribute of the decision tree, the input parameter with the highest information gain is chosen. The output of the decision tree-based classifier predicts the most appropriate vessel in a specific cluster to which emergency alerts can be sent from the shore [5]. The decision attribute in this work is the overall communication capability of the vessel labeled as good average and bad based on the environmental condition and network parameters of the vessel. The attribute with the highest information gain is recursively chosen as the splitting attribute for the node in the decision tree.

$$\text{Info}(D) = -\sum_{i=0}^{m} P_i \log_2(P_i) \tag{1}$$

$$\text{SplitInfo}_A(D) = -\sum_{j=1}^{v} \frac{|D_j|}{D} \log_2\left(\frac{|D_j|}{D}\right) \tag{2}$$

Equation (1) is used to compute the information of the whole system, where Info(D) is the information of the whole system. $SplitInfo_A(D)$ is the expected information gain due to a split with a particular attribute A is given by Eq. (2). From the input attributes, the gain ratio selects the attribute that maximizes the information [38].

$$\text{Gain Ratio}(A) = \frac{\text{Gain}(A)}{\text{SplitInfo}(A)} \tag{3}$$

Algorithm 2 explains the main steps in identifying the best node to pass alerts and messages during the time of adversity. DT*Classify()* is a decision tree-based classifier which classifies the input environmental and communication parameters. *Attribute_Selection_Method()* function finds out the attribute that best splits/classifies the nodes. Here, information-based attribute selection is used, and the attribute having the highest information is chosen for splitting the decision tree. *SentAlert()* sends alert messages to the nodes identified as good.

Algorithm 2 Algorithm for best node identification

```
DT Classify()
Preprocess the data set, D
analyze AL (Lat, Long, SST,Pressure,Wind Speed,SSI,Txn rate,Rxn rate )
Apply Attribute_selection_method(D, attribute_list)
Create a root node N
Label node N multiway splitting_criterion;
Classify nodes decision attributes(Good, Average, Bad)
List (D,I_good)
Sent Alert (I_good)
End
```

4 Results and Discussions

The dataset consists of the information regarding fishing vessels going for short fishing trips from Vizhinjam the south coast of Kerala 8° 22′ 21″ N, 76° 59′ 55″ E to as far as Manjeswaram port at North Kerala 12° 43′ 13.8″ N 74° 52′ 43.0″ E. The daily ocean state forecast from the Indian National Centre for Ocean Information Services (INCOIS) [39] is mapped to the corresponding GPS location data of the vessels in the sea. The signal strength, transmission rate, reception rate are received from the base station transceiver set up as part of the Micronet project. The attributes in the dataset and range are listed in Table 1.

Figure 3 shows the resultant clusters obtained after performing the FCM clustering algorithm on the dataset. The optimal sets of vessels are obtained through which the proper communication can be possible to proceed with the rescue services in no time.

Figure 4 presents FCM clustering result obtained with missing boats predicted location marked as M. The efficiency of the proposed algorithm is tested under a variety of conditions with different datasets. The algorithm was executed with a vessel dataset of over 2000 trajectories of about 100 fishing vessels available from

Table 1 Attribute list

Attribute name	Data type	Domain values
Vessel ID	Numerical	Unique ID
Latitude	Numerical	8° 22′–12° 43′
Longitude	Numerical	74° 52′-76° 59′
Sea surface temperature	Numerical	17–45 °C
Wind speed	Numerical	10–86 km/h
Vessel speed	Numerical	20–32 km/h
Vessel direction	Nominal	N, S, W, E
Signal strength	Numerical	−20 to −114 dBm

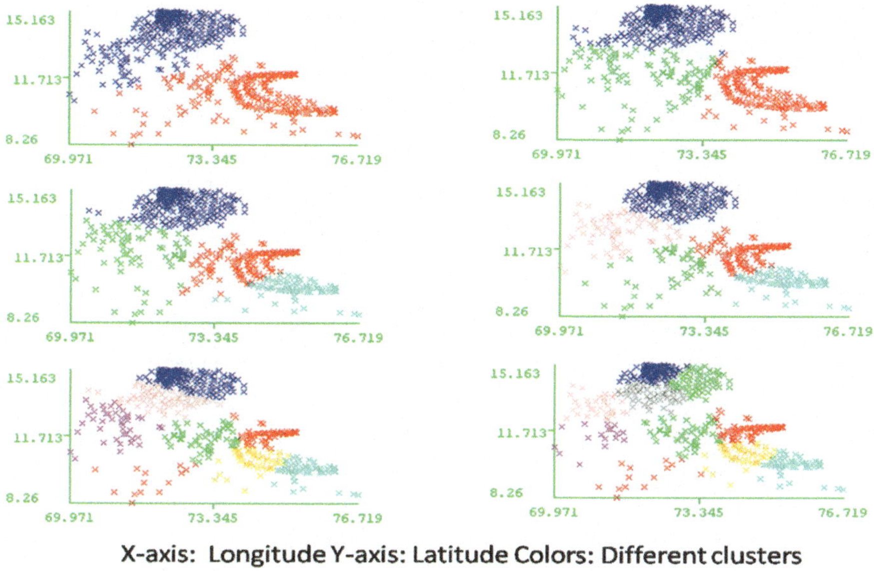

Fig. 3 Vessel clusters identified in deep sea based on missing details

Fig. 4 FCM clustering result with predicted location 'M'

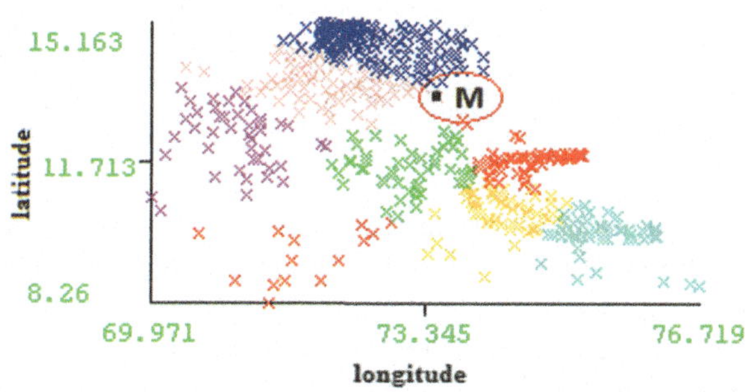

the Kerala coastal areas. Table 2 shows the sum of square errors (SSE) obtained for the clusters obtained using FCM clustering on different trajectory datasets with varying cluster size. It shows the intracluster cohesion showing the closeness of the points within a cluster. Lower SSE means high quality of clustering. A comparison with K-Means clustering is also evaluated, and results show that fuzzy clustering gives better results.

From Table 2, it is clear that the cluster quality is increasing with the increase in cluster size. From the clusters obtained through the FCM clustering algorithm, we can locate the vessel clusters near to the suspected region where the missing vessel is expected to be found. This is done by calculating the Euclidean distance between the cluster centroids with the location of these suspected regions. The nearest clusters are located, and information is conveyed to them to initiate the rescue operations.

Table 2 A comparison of the SSE Measures calculated for the clusters identified in the sea

Cluster size	Clustering algorithms	GPS dataset of fishing vessels in deep-sea fishing Kerala: Sum of Squared errors (SSE) measure			
		530 pts	850 pts	1044 pts	2000 pts
10	FCM	3.6782	7.6593	9.6163	14.7310
	K-means	3.9019	8.9769	10.1018	15.1381
9	FCM	4.1083	8.5952	10.2607	16.5724
	K-means	4.9812	10.0576	11.0897	16.3287
8	FCM	4.9006	9.5097	12.4484	19.0901
	K-means	5.1372	13.6781	12.9850	20.0781
7	FCM	5.6340	12.3027	14.6810	21.0261
	K-means	6.3978	14.0181	18.3174	24.0980
6	FCM	6.6962	13.0614	16.2086	24.2348
	K-means	8.3621	16.8732	20.8101	26.9107
5	FCM	7.9845	14.9962	17.9498	27.8912
	K-means	10.9980	19.3642	26.1814	29.5120
4	FCM	10.3787	17.1299	28.7325	35.7004
	K-means	15.4172	25.3851	30.1782	38.1900
3	FCM	16.0628	28.1587	33.4583	51.6468
	K-means	20.3129	35.6980	41.8901	60.5600
2	FCM	20.7647	38.3963	53.1084	77.9118
	K-means	25.3560	40.6715	55.9015	83.6117

In the optimal set of clusters, vessels with good signal strength and environmental conditions are detected and communication is passed to those vessels from the shore. The proposed vessel prediction algorithm paves way to an efficient means of finding the missing vessels in deep sea. It facilitates faster rescue services by intimating the nearest clusters of fishing vessels, thereby providing timely help instantly.

Table 3 shows the vessel location specification of the Kerala coastal region, which is considered for the experimental analysis in this work. Here the fishing vessels with vessel identification numbers V_ID49, V_ID51, V_ID30, V_ID39 from cluster 1 and V_ID25, V_ID10, V_ID76, V_ID58 are selected as the best nodes to which emergency communication can be sent from the shore to search the missing vessel. Figure 5 shows the decision tree-based induction obtained. It is found out from the result that parameter signal strength is having the highest information gain most of the time and is selected as the root node.

In order to compare the performance of the proposed algorithm, an experimental analysis and comparison is done with artificial neural network with the same dataset. A 10 input 3 output ANN with 3 intermediate layers trained with back propagation algorithm is used for our experiments. Table 4 shows the performance comparison result of the decision tree and artificial neural network-based classification, and it is found that generalized classification accuracy for decision tree-based classification is higher compared to neural network-based classification. The vessels in the clusters

Table 3 Vessel nodes identified in deep sea for communication alerts

	Vessel_ID	Longitude	Latitude	Wind speed	Sea surface temperature	SSI	Class
Cluster 1	V_ID49	71.05	12.88	20	36	−30	Good
	V_ID51	71.20	12.96	20	36	−35	Good
	V_ID30	71.31	12.48	20	36	−30	Good
	V_ID39	71.19	12.35	20	36	−35	Good
Cluster 2	V_ID25	73.59	13.22	20	36	−25	Good
	V_ID10	74.09	13.48	20	36	−30	Good
	V_ID76	73.82	13.27	20	36	−35	Good
	V_ID58	73.30	13.89	20	36	−25	Good

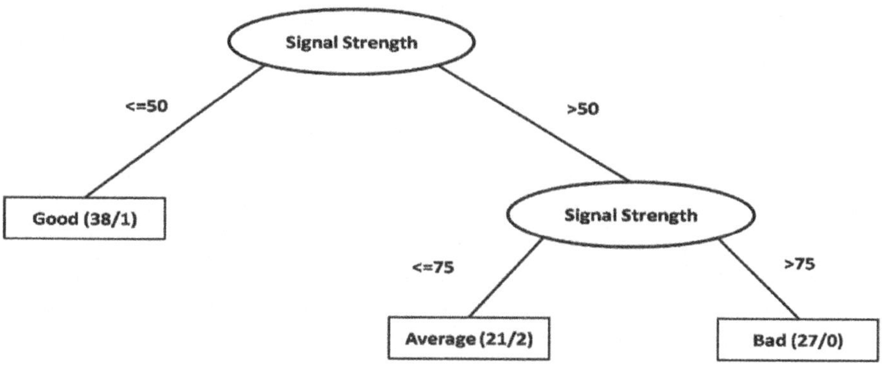

Fig. 5 Decision tree-based induction of vessels in deep sea

Table 4 Precision, recall and AUC for ANN and decision tree-based classifications

Algorithm	Class	Precision	Recall	AUC
Artificial neural network	Good	0.974	0.973	0.981
	Average	0.783	0.900	0.899
	Bad	0.960	0.828	0.956
Decision tree	Good	0.974	1.000	0.981
	Average	0.900	0.900	0.899
	Bad	0.964	0.931	0.971

are classified as good, average and bad, and the nodes which are classified as good with the highest probability are selected for emergency communication from the shore. The Receiver Operating Characteristics (ROC) plotted for the class good using decision tree algorithm is shown in Fig. 6. Area under curve for the characteristics is 0.98.

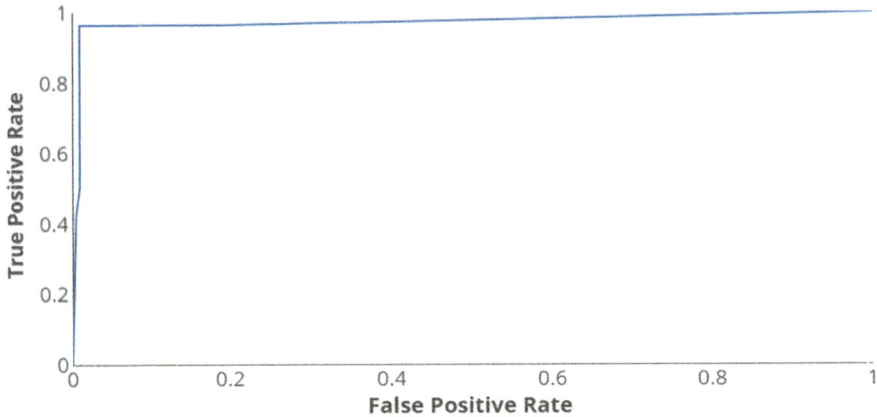

Fig. 6 ROC curve for decision tree-based classification for class: good AUC = 0.98

5 Conclusion

The proposed algorithm paves way to an efficient means of finding the missing vessels in deep sea during emergency situations by informing and communicating alert messages to the nearby vessels. The experimental results show that fuzzy C-means algorithm efficiently localizes the moving vessel clusters in the sea and decision tree-based classification can sharply identify the vessels to which timely communication alerts can be sent. Here, the possibilities of machine learning to track and locate missing vessels in maritime environment are analyzed. The proposed intelligent alert system helps in tracking and locating the isolated fishing vessels, people or any artifacts in marine when there is no communication possible with the search and rescue systems. This facilitates faster rescue services by intimating the nearest clusters of fishing vessels, thereby providing timely help. It reinforces the decision support systems with sensible knowledge in discovering user context, real-time mobility prediction of mobile nodes, using machine learning, especially to tackle emergencies like disaster alerts, missing vessel identification and so on.

Acknowledgements The work has been carried out as part of MICRONet project of Information Technology Research Academy (ITRA), Media Lab Asia, MeitY, Government of India.

References

1. Banerjee S, Bouzefrane S, Mühlethaler P (2017) Mobility prediction in vehicular networks: an approach through hybrid neural networks under uncertainty. In: International conference on mobile, secure, and programmable networking. Springer, Cham, pp 178–194
2. Muench A, DePiper GS, Demarest C (2017) On the precision of predicting fishing location using data from the vessel monitoring system (VMS). Can J Fish Aquat Sci 75(7):1036–1047

3. Shepperson JL, Hintzen NT, Szostek CL, Bell E, Murray LG, Kaiser MJ, O'Neill F (Handling editor) (2017) A comparison of VMS and AIS data: the effect of data coverage and vessel position recording frequency on estimates of fishing footprints. ICES J Mar Sci 75(3):988–998
4. Wang S, Wang S, Gao S, Yang W (2017) Daily ship traffic volume statistics and prediction based on automatic identification system data. In: 2017 9th International Conference on Intelligent Human-Machine Systems and Cybernetics (IHMSC), vol 2, IEEE, pp 149–154
5. Roy RA, Nair JP, Sherly E (2015) Decision tree based data classification for marine wireless communication. In: 2015 international conference on computing and network communications (CoCoNet), IEEE, 2015
6. Yavaş G et al (2005) A data mining approach for location prediction in mobile environments. Data Knowl Eng 54(2):121–146
7. Lu EH-C, Tseng VS (2009) Mining cluster-based mobile sequential patterns in location-based service environments. In: MDM'09 tenth international conference on mobile data management: systems, services and middleware, IEEE, 2009
8. Sheng P, Yin J (2018) Extracting shipping route patterns by trajectory clustering model based on automatic identification system data. Sustainability 10(7):2327
9. Cheng X, Xu J, Pei J, Liu J (2009) Hierarchical distributed data classification in wireless sensor networks. In: Proceedings of the IEEE international conference on Mobile Ad-hoc and Sensor Systems (MASS), 2009
10. Pathirana PN, Savkin AV, Jha S (2003) Mobility modelling and trajectory prediction for cellular networks with mobile base stations. In: Proceedings of the 4th ACM international symposium on mobile ad hoc networking & computing, ACM, pp 213–221
11. Zhang X, Yuan Z (2017) The GPS trajectory data research based on the intelligent traffic big data analysis platform. J Comput Methods Sci Eng 17(3):423–430
12. Andreas K, Krisp JM, Linfang D (2017) Detecting vehicle traffic patterns in urban environments using taxi trajectory intersection points Geo-Spatial. Inf Sci 20(4):333–344
13. Keler A, Krisp JM, Ding L (2017) Detecting travel time variations in urban road networks by taxi trajectory intersections. In: GISRUK 2017—25th GIS research UK conference, Manchester, UK, 18–21 Apr
14. Araki M, Kanamori R, Gong L, Morikawa T (2017) Impacts of seasonal factors on travel behavior: basic analysis of GPS trajectory data for 8 months, in serviceology for smart service system. Springer, Japan, pp 377–384
15. Chen W, Guo F, Wang F-Y (2015) A survey of traffic data visualization. IEEE Trans Intell Transp Syst 16(6):2970–2984
16. Chen BY, Yuan H, Li QQ, Shaw S-L, Lam WHK, Chen X (2016) Spatio temporal data model for network time geographic analysis in the era of big data. Int J Geogr Inf Sci 30:1041–1071
17. Dodge S, Weibel R, Ahearn SC, Buchin M, Miller JA (2016) Analysis of movement data. Int J Geogr Inf Sci 30:825–834. https://doi.org/10.1080/13658816.2015.1132424
18. Duruisseau M, Rouvoy R (2014) Automatic inference of roadmaps from raw mobility traces. Research report, RR-8585, INRIA. 19
19. Kong X, Li M, Li J, Tian K, Hu X, Xia F (2018) Copfun: an urban cooccurrence pattern mining scheme based on regional function discovery, World Wide Web. https://doi.org/10.1007/s11280-018-0578-x
20. Lloyd S (1982) Least squares quantization in PCM. IEEE Trans Inf Theory 28(2):129–137
21. Zhang T, Ramakrishnan R, Livny M (1996) BIRCH: an efficient data clustering method for very large databases. In: ACM Sigmod Record, ACM, vol 25, no 2, pp 103–114
22. Ester M, Kriegel HP, Sander J, Xu X (1996) A density-based algorithm for discovering clusters in large spatial databases with noise. In: Kdd, vol 96, no 34, pp 226–231
23. Ankerst M, Breunig MM, Kriegel HP, Sander J (1999) OPTICS: ordering points to identify the clustering structure. In: ACM Sigmod record, ACM, vol 28, no 2, pp 49–60
24. Wang W, Yang J, Muntz R (1997) STING: a statistical information grid approach to spatial data mining. In: VLDB, vol 97, pp 186–195
25. Gaffney S, Smyth P (1999) Trajectory clustering with mixtures of regression models. In: KDD, vol 99, pp 63–72

26. Lee JG, Han J, Whang KY (2007) Trajectory clustering: a partition-and-group framework. In: Proceedings of the 2007 ACM SIGMOD international conference on Management of data, ACM, pp 593–604
27. Li Z, Lee JG, Li X, Han J (2010) Incremental clustering for trajectories. In: International conference on database systems for advanced applications. Springer, Berlin, pp 32–46
28. Knorr EM, Ng RT, Tucakov V (2000) Distance-based outliers: algorithms and applications. VLDB J Int J Very Large Data Bases 8(3–4):237–253
29. Bomberger NA, Rhodes BJ, Seibert M, Waxman AM (2006) Associative learning of vessel motion patterns for maritime situation awareness. In: 2006 9th international conference on information fusion, IEEE, pp 1–8
30. Dahlbom A, Niklasson L (2007) Trajectory clustering for coastal surveillance. In: 2007 10th international conference on information fusion, IEEE, pp 1–8
31. Auslander B, Gupta KM, Aha DW (2011) A comparative evaluation of anomaly detection algorithms for maritime video surveillance. In: Sensors, and command, control, Communications, and Intelligence (C3I) technologies for homeland security and homeland defense X, International Society for Optics and Photonics, vol 8019, p 801907
32. Vespe M, Visentini I, Bryan K, Braca P (2012) Unsupervised learning of maritime traffic patterns for anomaly detection
33. Pallotta G, Vespe M, Bryan K (2013) Vessel pattern knowledge discovery from AIS data: a framework for anomaly detection and route prediction. Entropy 15(6):2218–2245
34. Liu B, de Souza EN, Matwin S, Sydow M (2014) Knowledge-based clustering of ship trajectories using density-based approach. In: 2014 IEEE international conference on big data (big data), IEEE, pp 603–608
35. Lei PR (2016) A framework for anomaly detection in maritime trajectory behavior. Knowl Inf Syst 47(1):189–214
36. Zhen R, Jin Y, Hu Q, Shao Z, Nikitakos N (2017) Maritime anomaly detection within coastal waters based on vessel trajectory clustering and Naïve Bayes Classifier. J Navig 70(3):648–670
37. Chitra MT, Sherly E (2016) Trajectory clustering and mobility pattern identification of missing vessels in deep sea. In: International conference on robotics and automation (IETE ICRA 2016), Trivandrum, Kerala, Feb 2016
38. Quinlan JR (1986) Induction of decision trees. Mach Learn 1(1):81–106
39. https://incois.gov.in/portal/osf/location_specific.jsp

A Study of CNN Architectures over Two Hand Indian Sign Language Dataset

Vaidehi Sharma, Mohita Jaiswal, Abhishek Sharma, and Raghuvir Tomar

Abstract Sign language is a nonverbal way of communication with the speech-impaired and hearing-disabled people. Hand gestures play a significant role in establishing the connection between humans and machines. Gesture recognition poses a serious challenge to the system due to improper light and background conditions. Recognition of gestures using two hands is more difficult than using one hand gestures. In the present work, we have used Two Hand Indian Sign Language (THISL) dataset and have added two more gestures to it. The modified dataset consists of approximately 12,600 images which split into 80% for training set and the remaining 20% for the test set. The aim is to validate THISL dataset on various convolutional neural network models and analyze the model's performance in terms of accuracy and loss, computational time, and memory usage.

Keywords Indian Sign Language · Deep learning · Hand sign recognition · Computer vision · Convolutional neural networks (CNN) · Machine learning (ML) · Artificial neural network (ANN) · Deep learning (DL)

1 Introduction

Hand sign language recognition has been gaining significant attention in the area of computer vision research [1–10]. This technique serves as a medium of communication for hearing-disabled and vocally impaired people. Since most people do not

V. Sharma (✉) · M. Jaiswal · A. Sharma · R. Tomar
The LNM Institute of Information Technology, Jaipur, Rajasthan 302031, India
e-mail: vaidehisharma444@gmail.com

M. Jaiswal
e-mail: mohitajaiswal166@gmail.com

A. Sharma
e-mail: abhisheksharma@lnmiit.ac.in

R. Tomar
e-mail: rtomar@lnmiit.ac.in

S. M. Thampi et al. (eds.), *Applied Soft Computing and Communication Networks*, Lecture Notes in Networks and Systems 125,
https://doi.org/10.1007/978-981-15-3852-0_14

know about sign language, so there exists a need for a system that translates sign language into regional language automatically.

Most sign languages include static as well as dynamic sign gestures. The static gestures are those which are made using one hand, and the dynamic gestures are those which are made using two hands [6]. Additionally, there are some gestures that require the coordination of both hand gestures and facial expressions. Consequently, the task of gesture recognition software is made difficult because of the difficulty in identifying the proper word. The task difficulty is further increased by the fact that while making gestures, the two hands may not move with the same speed. In India, professionals believe that there is an acute shortage of special schools for deaf and dumb people. Therefore, there are very few people who speak and know about sign language. In India, the sign language of a particular region depends on the culture and the spoken language of that particular region.

Generally speaking, a sign language comprises the vocabulary of gestures in the same way as spoken language contains vocabulary of words. The most commonly used datasets that exists for sign languages are MSR Gesture3D [11], American Sign Language Lexicon Video Dataset [8], LTI-Gesture Database [9], Auslan dataset [10], DEVISIGN Chinese Sign Language dataset [12], RWTH German Finger spelling [13], and Indian Sign Language dataset [14].

1.1 Deep Learning

With the evolution in the area of computer vision research, deep learning has become an attractive choice for solving complex problems requiring large computational capability. According to a recent survey [11], hand sign recognition has become one of the successful applications in the field of deep learning. The advantage of using DL models for image classification is that it does not need preprocessing of the images with blob detectors to extract relevant features and then use classifier for classification. The CNN architecture solves the problem of feature extraction and classification and also gives better performance over image classification.

A methodology has been proposed in [15] based on dynamic time warping to overcome the limitation of glove-based traditional approach. This algorithm analyzes variations in the hand positions along with facial expressions. The work reported in [16] introduces the dataset of mobile selfie Indian Sign Language and evaluates it on different CNN architectures. Another implementation is presented in [17] based on hand sign digits recognition and converted text into Bangla speech. The deep learning methodology reported in [18] is used for recognizing an Indian Sign Language (ISL) static alphabet from binary silhouette of signer hand region. The work reported in [19] involves recognition of Indian Sign Language techniques by finger spelling. It makes use of ANN and techniques of digital image processing for recognition of gestures. The steps involved in this paper are acquisition, extraction of features, preprocessing of the data, and classification. An automation system is proposed in [20] to recognize ISL in complex backgrounds that involve natural language processing, pattern recognition, and computer vision. For this experiment, webcam is used for the image acquisition, frames of the video are preprocessed and skin segmented, and lastly, feature extraction and classification are used to recognize the respective signs.

1.2 Outline

In the present work, we have collected Two Hand Indian Sign Language (THISL) dataset [19] consisting of 26 gestures including English alphabets. We have added two more gestures of 'delete' and 'space' to the dataset. So, the dataset which we have used consists of 28 gestures. The performance of different CNN architectures has been measured and compared over THISL dataset in terms of loss and accuracy.

The structure of paper is systematized as follows. Section 2 introduces procedure used for the collection of THISL dataset. In Sect. 3, several CNN architectures are discussed. Section 4 presents the procedure used for implementation of various CNN models. In Sect. 5, the detailed analysis of results and a discussion on the performance of various CNN models are presented. In the final Sect. 6, conclusions are drawn and the scope of future work is presented.

2 Dataset Collection

Databases handling two hand gestures and suitable for Indian Sign Language are generally not available. THISL dataset that we have used is quite suitable for handling two hand gestures and converting the gestures into a readable form.

The THISL dataset used in this work is shown in Fig. 1. Compared to the previously reported dataset, we have added two more gestures, thereby making the total no. of gestures to be 28. This modified dataset was manually collected using Digital Single-Lens Reflex (DSLR) camera. In this dataset, 28 gestures were recorded at the natural position where they were actually performed by both hands. Out of the 28 gestures, 26 gestures comprise English alphabets 'A' to 'Z' and the remaining two gestures are 'space' and 'delete.' Each class of gesture has a total of 450 images. The combined dataset contains 12,600 images.

3 CNN Architectures

In the areas of ML, ANN performs well for image and audio classification [22]. Similarly, for deep learning applications, researchers have been paying more attention to the convolutional neural network (CNN) so as to improve the performance and make the system suitable for real-time practical situations [23].

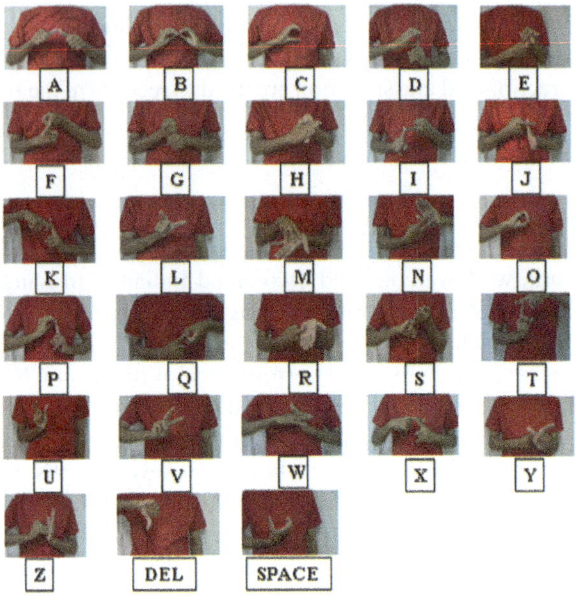

Fig. 1 THISL dataset [21]

The classical CNN architectures usually consist of different layers, for example, convolutional layer, fully connected layer, and pooling layer. The organization of CNN constituents portrays a vital role in designing new architectures, thus accomplishing enhanced performance of the CNN models. This section briefly describes the mathematics behind the CNN architectures, like choosing of activation functions, dimensions of the output matrix, etc.

Convolutinal layer: It is composed of set of neurons, also known as kernels. With small region of the image, and that particular region is known as receptive field. Suppose we have $N \times N$ neuron layer and filter of size $m \times m$, then convolutional layer size will be $(N - m + 1), (N - m + 1)$.

$$Z_k^l = (I_{x,y} * K_k^l) \tag{1}$$

$I_{x,y}$ represents an input image, and l denotes the l-th convolutional kernel of k-th layer. Basically, convolution is a technique, in which a matrix of numbers, that is called filter or a kernel, is passed through the given input image and renovate it based on the filter value. This layer helps to extract features from an image.

Pooling layer: The use of pooling operations helps to extract combinations of features. In these layers, input image is divided into separate regions, and then, some operations are achieved on each part of the experimented image. For example, in max pooling, a maximum value is selected from each region and puts in the corresponding output matrix [24].

$$P_l = f_p(F_{x,y}^l) \quad (2)$$

P_l represents the l-th output feature map. $F_{x,y}^l$ shows the l-th input feature map. $f_p(.)$ denotes the type of pooling operation. Various pooling operations include max pooling, average pooling, L2, etc.

Activation Functions: Activation functions are decision functions that help in learning the complex patterns. Choosing the appropriate activation function leads to accelerate the learning process. Convolved feature map is defined as:

$$A_l^k = f_a(Z_l^k) \quad (3)$$

Z_l^k represents the output of convolutional operations that is assigned to activation function. f_a adds nonlinearity and returns the transformed output. Different types of activations such as sigmoid, Tanh, ReLU, and Leaky ReLU are used to add nonlinear combinations of features. The dimensions of the output matrix after adding padding and strides can be calculated using

$$n_{\text{out}} = [\frac{n_{\text{in}} + 2p - f}{s} + 1] \quad (4)$$

n_{in} represents the input image size, p is for padding of the zeros, f denotes the filter size, and s stands for strides.

The main idea behind the CNN is to reduce network structure complexity by sharing sensory fields and weights, thus making it faster for parallel computations. Some of the CNN architectures used for image classification task are discussed in this section.

3.1 LeNet

LeNet was first suggested in the 1990s by Yann LeCun and was given the name LeNet-5 [25]. MNIST dataset [26] was used to evaluate the performance of the architecture. The basic configuration of LeNet-5 network architecture is that it uses filter size of 5×5 and a stride of 1 in convolutional layers. The input layer consists of grayscale images of size 32×32, followed by convolutional (CONV), pooling layers alternately, and fully connected layers (FC).

3.2 AlexNet

AlexNet was proposed in 2012 [27]. It was the major breakthrough in CNN models and has won ImageNet Large Scale Visual Recognition Challenge (ILSVRC).

This model is deeper and wider than the LeNet model. Alexnet architecture comprises input layer with 227×227 RGB images, 5 CONV layers, dropout layers, and 3 FC layers. These eight layers are united with max pooling and ReLU activations which show improved training performance over tanh activation and sigmoid activation functions. In order to overcome the problem of overfitting, local response normalization (LRN) and dropout are introduced [28].

3.3 GoogleNet

GoogleNet or inception network has won the ILSVRC 2014 classification challenge. This architecture was introduced by Szegedy et al. [29]. This network aims at reducing computational complexity by using dimensionality reduction. Initially, in the naive version of the architecture, the inception layer consists of variable receptive field created by different kernel sizes. In this architecture, a number of 1×1 filters are used before the 3×3 and 5×5 convolutions in the reduction layer. The final inception network is improved by reducing its dimensions and increasing the computational efficiency.

3.4 Residual Network (ResNet)

ResNet has won the ILSVRC 2015 classification challenge. This architecture was proposed by He [30] with the intent to ease the training by increasing depth of network with skip connections and to reduce the network complexity compared to the classical CNN. Recent research [29, 31] reveals the importance of increasing depth of the network. The plain network as shown in Fig. 2 (Left) with deeper networks tends to degrade the performance due to vanishing or exploding gradients. Here, the output of layer is denoted by $F(x)$, i.e, unreferenced mapping. In order to solve this problem, a deep residual learning network with shortcut connections by skipping one or more layers is employed as shown in Fig. 2 (right). Here, the output is denoted by underlying mapping $H(x) := F(x) + x$. Realization of block connections in residual network is done by summing operations, and the curved arrows show the skip connections to summing operation [32]. ResNet is constructed with several different layers: 18, 34, 50, 101, 152, and even more than 1000. Out of all of these, most widespread is ResNet-50, that comprises 49 CONV layers and a single FC layer.

Fig. 2 Plain network (Left) and residual network (Right)

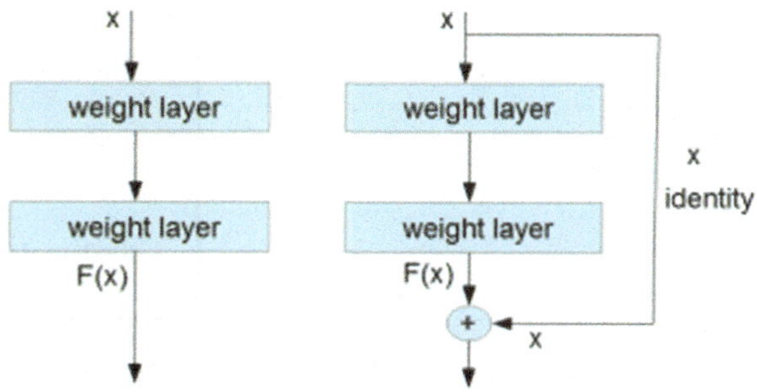

4 System Implementation

In this section, THISL dataset is used to implement the various models on different CNN architectures. The procedure used for building a CNN model over training images is also discussed.

4.1 CNN Training Model

A THISL dataset consists of 12,600 RGB images of 28 gestures, where 26 gestures are representing English alphabets and the remaining two gestures represent 'delete' and 'space.' The images of gestures were collected for the experiments under simple background. In the experiments, 80% of the image data belongs to training set and remaining 20% data are present in the test set. The data is preprocessed according to the standard architecture requirements. The images are resized into 224×224 as input images for AlexNet, ResNet-50, and GoogleNet models, and the images are resized into 32×32 for the LeNet [26]. The procedure of training CNN model is shown in Fig. 3. Primarily, the raw data is split into train and test images, and then, the images are resized conferring to the CNN architecture and as per the number of classes, and then, the data is labeled. After the CNN model is built, loss function is measured and optimizer is used to minimize the loss function. Finally, the predictions are made based on the true label and predicted label.

4.2 Network Training Phase

The four different CNN models are trained on the 10k training images and are validated on 2k test images. They are implemented by Keras taking Tensorflow as backened. While training CNN model, the optimizer used is stochastic gradient descent (SGD) with batch size of 256 and categorical cross-entropy is used as the loss function. The learning rate starts from 0.001 and improves the network performance

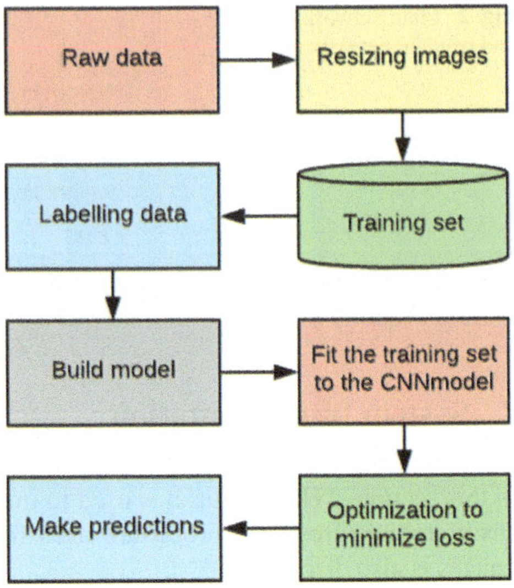

Fig. 3 CNN training model

during the training process. The model is trained for nearly 50 epochs to observe the precise stabilizations for test errors. The NVIDIA Quadro K4200 GPU with 400GB memory is used to train the CNN model.

5 Results and Analysis

5.1 Experimental Results

THISL dataset was benchmarked on various CNN architectures of deep learning by tuning the hyperparameters such as learning rate and choice of activation function. The models were trained nearly on 50 epochs with batch size of 256 using Python 3. The curve of training accuracy and test accuracy versus epoch and the curve of training loss and validation loss versus epoch are shown for each CNN model in Figs. 4, 5, 6, and 7.

As we can see from Fig. 4(a), (b), with increasing number of epochs, the LeNet model test accuracy reaches approx to 88%, and on the other hand, the test loss continuously decreases to 0.5.

Figure 5(a), (b) shows the performance of AlexNet model. With the increase in no of epochs, the test accuracy is increased to 98% and then becomes constant throughout the training, and the value of test loss is minimized to 0.01.

From Fig. 6(a), (b), it is concluded that GoogleNet model test accuracy increases after a sudden fall at some epochs in between and then becomes constant throughout the training and reaches to 99%, The test loss is decreased approx to 0.011.

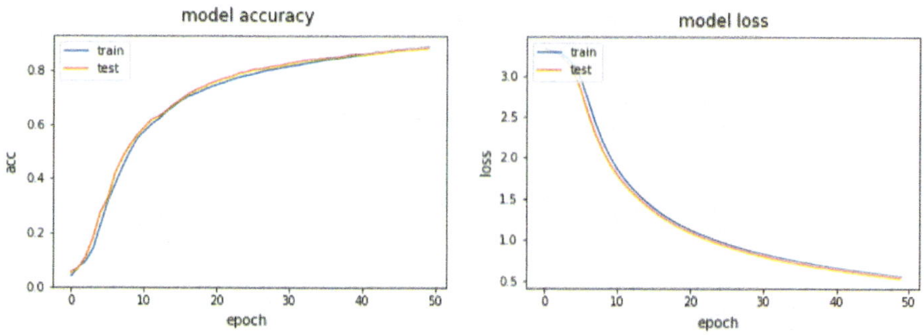

(a) Training and Test accuracy (b) Training and Test loss

Fig. 4 Training and test performance of LeNet model trained on THISL dataset

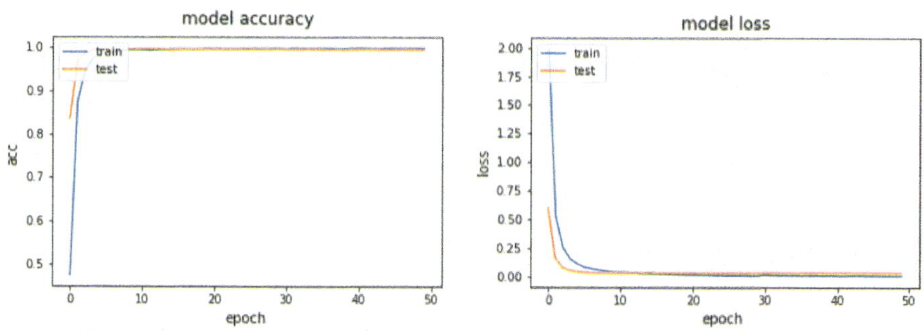

(a) Training and Test accuracy (b) Training and Test loss

Fig. 5 Training and test performance of AlexNet model trained on THISL dataset

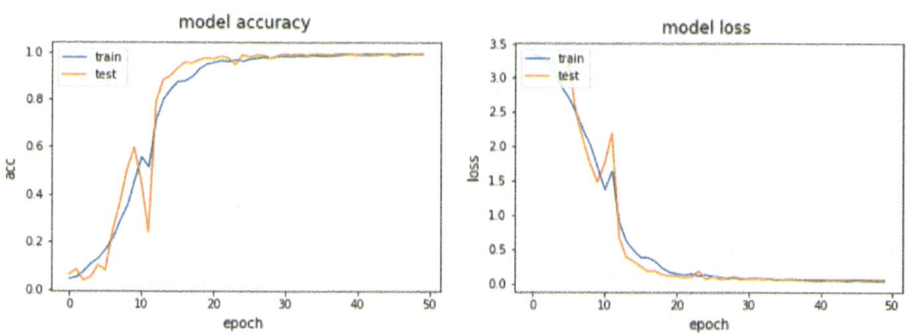

(a) Training and Test accuracy (b) Training and Test loss

Fig. 6 Training and test performance of GoogleNet model trained on THISL dataset

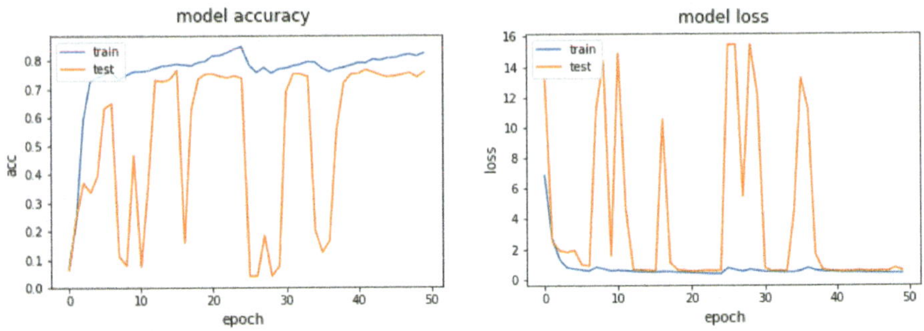

(a) Training and Test accuracy

(b) Training and Test loss

Fig. 7 Training and test performance of ResNet-50 model trained on THISL dataset

Figure 7(a), (b) shows the performance of ResNet-50 model. Due to the problem of overfitting, the curves of test loss and accuracy gradually increase and decrease throughout the training process.

5.2 *Performance Analysis*

The performance of the trained CNN architecture is analyzed on the basis of the following parameters, test loss and accuracy, computational time of training model, and memory usage. Table 1 shows benchmarking various CNN classification models over the dataset.

The computational time for training procedure of LeNet is approximately 10 min. The GoogleNet lasts for 1 day, and for AlexNet and ResNet-50, the training time is four times and eight times longer than GoogleNet architecture, respectively. The usage of memory of each architecture mentioned in Table 1 is checked through the memory profiler which gives total memory consumption of python code.

Table 1 Performance analysis of various CNN architectures

CNN Architectures	Test loss	Test accuracy(%)	Memory usage(MB)
LeNet	0.543	88.5	0.54
AlexNet	0.014	98.9	225
GoogleNet	0.011	99.6	24.3
ResNet-50	0.491	81.3	294

6 Conclusions and Scope of Future Work

In this work, the performance of CNN architectures is analyzed in terms of metrics of loss and accuracy, computational time, and memory consumption over double hand gestures dataset consisting of 28 classes. Prior to the training stage, the input images are resized according to the requirement of architectures. From the performance analysis, we conclude that the GoogleNet is performing better as compared to other architectures with 99.6% test accuracy and 0.011 value of test loss, and also, in terms of time taken during training process and memory consumption, the GoogleNet is more suitable in performing double hand gesture classification on Indian Sign Language dataset. In the future, we are planning to extend our work by applying data augmentation technique to increase the number of images of gestures in each class. The aim is to avoid the overfitting of data and to improve the efficiency of CNN architectures.

Acknowledgements We thank Ms. Runa Roka from Noida Deaf Society for helping us in the collection of THISL dataset and for their expert opinions. This work is supported by Department of Science and Technology (DST), India, under project name Sign Language to Regional Language Converter (SLRLC) with project number SEED/TIDE/063/2016.

References

1. Anwar S, Sinha SK, Vivek S, Ashank V (2019) Hand gesture recognition: a survey. In: Nanoelectronics, circuits and communication systems. Springer, Berlin, pp 365–371
2. Ghotkar AS, Khatal R, Khupase S, Asati S, Hadap M (2012) Hand gesture recognition for Indian sign language. In: 2012 international conference on computer communication and informatics. IEEE, pp 1–4
3. Chaudhary A, Raheja JL, Das K, Raheja S (2013) Intelligent approaches to interact with machines using hand gesture recognition in natural way: a survey. arXiv preprint. arXiv:1303.2292
4. Molchanov P, Gupta S, Kim K, Kautz J (2015) Hand gesture recognition with 3D convolutional neural networks. In: Proceedings of the IEEE conference on computer vision and pattern recognition workshops, pp 1–7
5. Cheok MJ, OmarZ, Jaward MH (2019) A review of hand gesture and sign language recognition techniques. Int J Mach Learn Cybern 10(1):131–153
6. Ghotkar AS, Kharate GK (2014) Study of vision based hand gesture recognition using Indian sign language. Computer 55:56
7. Wang J, Liu Z, Chorowski J, Chen Z, Wu Y (2012) Robust 3D action recognition with random occupancy patterns. In: European conference on computer vision. Springer, Berlin, pp 872–885
8. Athitsos V, Neidle C, Sclaroff S, Nash J, Stefan A, Yuan Q, Thangali A (2008) The American sign language lexicon video dataset. In: 2008 IEEE computer society conference on computer vision and pattern recognition workshops. IEEE, pp 1–8
9. Ney IH, Dreuw P, Seidl T, Keysers D (2005) Appearance-based gesture recognition
10. SignBank and Auslan. http://www.auslan.org.au/about/dictionary/
11. Wang Jindong, Chen Yiqiang, Hao Shuji, Peng Xiaohui, Lisha Hu (2019) Deep learning for sensor-based activity recognition: a survey. Pattern Recogn Lett 119:3–11

12. Chai X, Wang H, Chen X (2014) The devisign large vocabulary of Chinese sign language database and baseline evaluations. Technical report VIPL-TR-14-SLR-001. Key Lab of Intelligent Information Processing of Chinese Academy of Sciences (CAS), Institute of Computing Technology, CAS
13. Rwth german fingerspelling database. https://www-i6.informatik.rwth-aachen.de/aslr/fingerspelling.php
14. Kumar P, Gauba H, Roy PP, Dogra DP (2017) A multimodal framework for sensor based sign language recognition. Neurocomputing 259:21–38
15. Ahmed W, Chanda K, Mitra S (2016) Vision based hand gesture recognition using dynamic time warping for Indian sign language. In: 2016 International Conference on Information Science (ICIS). IEEE, pp 120–125
16. Anantha Rao G, Syamala K, Kishore PVV, Sastry ASCS (2018) Deep convolutional neural networks for sign language recognition. In: 2018 conference on Signal Processing and Communication Engineering Systems (SPACES). IEEE, pp 194–197
17. Ahmed S, Islam M, Hassan J, Ahmed MU, Ferdosi BJ, Saha S, Shopon M et al (2019) Hand sign to bangla speech: a deep learning in vision based system for recognizing hand sign digits and generating bangla speech. arXiv preprint arXiv:1901.05613
18. Sruthi CJ, Lijiya A (2019) Indian sign language recognition using ANN and SVM classifiers. In: 2019 International Conference on Communication and Signal Processing (ICCSP). IEEE, pp 0596–0600
19. Ekbote J, Joshi M (2017) Indian sign language recognition using ANN and SVM classifiers, pp 1–5
20. Reshna S, Jayaraju M (2017) Spotting and recognition of hand gesture for Indian sign language recognition system with skin segmentation and SVM, pp 386–390
21. Mangamuri LST, Jain L, Sharma A (2019) Two hand Indian sign language dataset for benchmarking classification models of machine learning. Accepted
22. Kiranyaz S, Avci O, Abdeljaber O, Ince T, Gabbouj M, Inman DJ (2019) 1D convolutional neural networks and applications: a survey. arXiv preprint arXiv:1905.03554
23. Wang Z, Liu K, Li J, Zhu Y, Zhang Y (2019) Various frameworks and libraries of machine learning and deep learning: a survey. In: Archives of computational methods in engineering, pp 1–24
24. Ren S, He K, Zhang X, Sun J (2015) Spatial pyramid pooling in deep convolutional networks for visual recognition. IEEE Trans Pattern Anal Mach Intel 37(9):1904–1916, 52pp
25. LeCun Yann, Bottou Léon, Bengio Y, Haffner P et al (1998) Gradient-based learning applied to document recognition. Proc IEEE 86(11):2278–2324
26. LeCun Y, Cortes C (2010) Mnist handwritten digit database. http://yann.lecun.com/exdb/mnist/
27. Krizhevsky A, Sutskever I, Hinton GE (2012) Imagenet classification with deep convolutional neural networks. In: Advances in neural information processing systems, pp 1097–1105
28. Alom MZ, Taha TM, Yakopcic C, Westberg S, Sidike P, Nasrin MS, Van Esesn BC, Awwal AAS, Asari VK (2018) The history began from alexnet: a comprehensive survey on deep learning approaches. arXiv preprint arXiv:1803.01164
29. Szegedy C, Liu W, Jia Y, Sermanet P, Reed S, Anguelov D, Erhan D, Vanhoucke V, Rabinovich A (2015) Going deeper with convolutions. In: Proceedings of the IEEE conference on computer vision and pattern recognition, pp 1–9
30. He K, Zhang X, Ren S, Sun J (2016) Deep residual learning for image recognition. In: Proceedings of the IEEE conference on computer vision and pattern recognition, pp 770–778
31. Simonyan K, Zisserman A (2014) Very deep convolutional networks for large-scale image recognition. arXiv preprint arXiv:1409.1556
32. He K, Sun J (2015) Convolutional neural networks at constrained time cost. In: Proceedings of the IEEE conference on computer vision and pattern recognition, pp 5353–5360

Periocular Recognition Under Unconstrained Conditions Using CNN-Based Super-Resolution

Vineetha Mary Ipe and Tony Thomas

Abstract A major challenge of the prevailing biometric systems is the short range of trait capture. Relaxing the range constraints imposed on the subjects arouses advanced challenges in the quality of image and resolution. In such scenarios, the lack of quality of acquired biometric information can be addressed using super-resolution, a technique of generating high-resolution images from low resolution counterparts. Procurement of the periocular images requires less cooperation of subjects compared to other ocular biometrics, thereby emerging as a reliable trait for unconstrained biometrics. For our work, UBIRIS v.2 is used which provides the relevant data for less constrained biometrics. The best rank-1 accuracy reported for periocular recognition with this database is 87.62%, and such works rely on the local and global feature descriptors. On this account, we cannot possibly arrive at the conclusion that these attributes in the literature are the best descriptors for the periocular region. One possible solution to achieve better recognition performance is to employ the latest trends in deep learning. Accordingly, we propose to apply deep learning-based super-resolution technique to the periocular images for improved identification efficiency. Our research work is novel in the prospect that this is the first study for periocular recognition applying CNN-based super-resolution. The off-the-shelf CNN features used in our work give improved rank-1 accuracy of 91.47%.

1 Introduction

Periocular recognition is the automated process of identifying individuals using periocular images. Periocular is the facial region around the eye including eyelids, eyelashes, and eyebrows [1]. The significance of periocular biometrics lies in the fact

V. M. Ipe (✉) · T. Thomas
Indian Institute of Information Technology and Management-Kerala, Technopark, Trivandrum, India
e-mail: vineetha.mary@iiitmk.ac.in

T. Thomas
e-mail: tony.thomas@iiitmk.ac.in

S. M. Thampi et al. (eds.), *Applied Soft Computing and Communication Networks*, Lecture Notes in Networks and Systems 125,
https://doi.org/10.1007/978-981-15-3852-0_15

that it represents a trade-off between using face and iris [2]. The distinctive markings of the iris, sclerotic blood vessels, and skin texture attribute to the uniqueness of the periocular (periphery of ocular) region for each individual [3].

A major challenge in current biometric systems is the range of image capture, staying motionless around 60 cm from iris sensors or a few meters away from face cameras. This distance constraint can be relaxed to a certain extent by making use of the periocular biometrics. Acquisition of the periocular images requires less subject cooperation compared to other ocular biometrics thereby emerging as a promising trait for unconstrained biometrics.

Resolution of an image represents the details encompassed in it. High-resolution implies greater density of pixels and hence rich visual details. In most applications, high-resolution (HR) images are often required since they portray lucid pictorial details for machine learning [4]. But, low resolution (LR) images are ubiquitous in real-world scenarios such as surveillance owing to the cheap low resolution image capture devices, optics employed in the capture, and high separation between participants and camera. Resolution can be improved by making use of high precision optics and long focal lenses. However, employing such advanced hardware solutions is not always feasible or economical since high cost is incurred.

Adopting software techniques for increasing the resolution of an image such as interpolation (bicubic, bilinear, and nearest neighbor) [5] results in blurred and blocky images (staircase-effect). Therefore, there is a requirement to consider another approach to overcome these shortcomings. In such cases, super-resolution (SR) [4] can be used to overcome the inadequacy of resolution in the biometric systems. Super-resolution techniques boost the quality and resolution of the captured images, thus relaxing the constraints on the prevailing biometric systems.

For our experiments, UBIRIS v.2 [6] is used which contains ocular images captured in visible wavelength at a distance and on the move, thus providing the appropriate data for less constrained biometrics. The best rank-1 accuracy reported for periocular recognition with UBIRIS v.2 is 87.62% [3], and such works rely on the feature descriptors [1] both global and local. Taking this into account, we cannot possibly arrive at the conclusion that these features in the literature are the best descriptors for the periocular region. One possible solution to achieve better recognition performance is to exploit the latest trends in deep learning. Latest advancements in deep learning suggest that the generic features extracted from pretrained convolutional neural networks (CNNs) can characterize intricate images such as that of periocular [7]. Additionally, research suggests that more the resolution of biometric images, better is the recognition accuracy [8]. In recent years, deep CNNs have evolved and are showing a significant role in enhancing the resolution of images. Accordingly, we propose to apply deep learning-based super-resolution technique namely VDSR to the periocular images so as to get improved recognition performance. Our research work is novel in the aspect that this is the first study for periocular recognition applying CNN-based super-resolution. The off-the-shelf CNN features used in our work provide an improved 91.47% rank-1 accuracy. We also analyze the periocular recognition performance with and without super-resolution. Our results show that there is a significant hike in the recognition performance by the application of SR in periocular images.

2 Related Works

The pioneering work in the field of periocular biometrics was reported in the literature by Park et al. in 2009 [9]. They made a feasibility study of whether periocular can be used as a biometric trait for recognition and confirmed the same [2]. Park et al. used several methods for feature extraction, such as scale-invariant feature transform (SIFT) [10], histograms of oriented gradients (HOG) [11], and local binary patterns (LBP) [12]. These works have paved the way to continuously enhance the effectiveness of periocular recognition methods.

Deep learning approaches especially CNNs have gained boundless popularity for pattern analysis and computer vision tasks. However, surveys on periocular biometrics [1] suggest that few studies have taken advantage of deep learning methods for improving periocular recognition rates. Zhao et al. [13] proposed a semantic-assisted convolutional neural network (SCNN) method which uses semantic information apart from learning identities. This technique involves learning from scratch which is computationally expensive and does not have a significant boost in the recognition performance. Proenca et al. [14] proposed a deep learning framework for periocular recognition without iris and sclera (Deep-PRWIS). Though this is an attempt to highlight the importance of periocular region alone, research suggests that use of entire periocular region without masking provides better recognition performance [2], which is the crux of the biometrics. Kevin et al. [7] performed periocular recognition using off-the-shelf CNN features. Compared to the conventional periocular features such as LBP, HOG, and SIFT, they show an equal error rate reduction up to 0.4. Considering the pace at which research in deep learning is progressing, we can anticipate much more precise and fast algorithms.

State-of-the-art works are interested in relaxing the imaging constraints imposed during the acquisition of biometric traits. Some of those works include fusing periocular and iris features together. However, a great delay is experienced in feature extraction methods employed [15]. Also, relaxing the distance constraints imposed on the participants raises new obstacles in resolution and image quality. The lack of quality of acquired biometric information can be addressed using super-resolution, a technique of creating high-resolution images from low resolution counterparts. Very few works are reported in literature for super-resolution in periocular images. Tapia et al. [16] presented a super-resolution approach for gender classification from mobile periocular iris images. Reddy et al. [17] presented a multiframe super-resolution for ocular biometrics. However, this work cannot be considered as a super-resolution on periocular biometrics since the images used are devoid of eyebrows, which is an important aspect in deciding the accuracy of periocular recognition [2].

As mentioned in the previous section, we use the UBIRIS v.2 for periocular recognition. This database is highly useful for researchers focusing on less constrained biometrics. Bakshi et al. [3] proposed a phase-intensive local pattern (PILP) for periocular recognition under visible spectrum, which is reported to give the best rank-1 accuracy for periocular recognition in the literature. But the value obtained is not lucrative enough to accept these features as the best descriptors for periocular images, and there are restrictions to enhance the performance since it is cumbersome to arrive at the optimal features. One possible remedy for performance improvement is to incorporate super-resolution and deep learning techniques. Based on these facts, we apply super-resolution to the periocular images and extract the off-the-shelf CNN features from Alexnet so as to train an SVM-based image classifier.

3 Proposed SR-Based Periocular Recognition

The framework for periocular recognition is as shown in Fig. 1. This recognition framework is one of its kind whereby periocular images undergo CNN-based super-resolution before being subjected to feature extraction. This kind of super-resolution is known as Very Deep Super-Resolution (VDSR), which is very efficient in improving the resolution of images with reliabile reconstruction.

3.1 Periocular Segmentation

The periocular region is segmented from these images by modifying the Viola–Jones algorithm [18]. Such a segmentation routine is precise ensuring the incorporation of vital periocular information such as eye, eyebrow, and surrounding skin texture. After segmentation, the periocular images belonging to distinct individuals are grouped together.

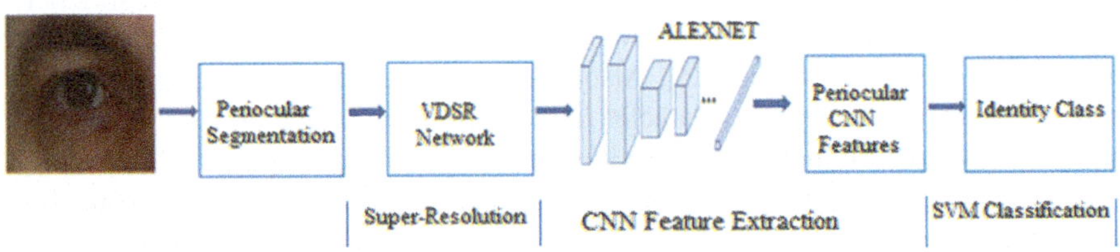

Fig. 1 Experimental framework

3.2 Super-Resolution

Super-resolution (SR) is the process of creating high-resolution images from low resolution images. The periocular images extracted from the UBIRIS v.2 is subjected to SR, and the high-resolution images so obtained are beneficial in improving the recognition accuracy as already pointed out. SR can be realized using deep learning [4], and this has shown significant reconstruction efficiency. Some of the deep learning approaches for SR are super-resolution convolutional neural network (SRCNN), sparse coding super-resolution (SCN), and learned iterative shrinkage thresholding algorithm (LISTA). Compared to these, very deep super-resolution (VDSR) obtains the best results with least testing time. Although super-resolution generative adversarial network (SRGAN) can produce photo-realistic performance, huge computational cost and time are incurred in training. Additionally, the output from SRGAN may look photo-realistic, but it can produce fake patterns as well (not real). Furthermore, the peak signal to noise ratio (PSNR) and structural similarity (SSIM) index are much higher for VDSR than SRGAN. The comparison of SRGAN and VDSR in terms of PSNR and SSIM for the benchmark datasets Set 5 and Set 14 [19, 20] is summarized in Table 1. Taking these into consideration, we opt to perform VDSR on the periocular images.

VDSR, which uses deep residual learning, is a 20 layers deep CNN model stacking 18 convolutional layers with rectified linear unit (ReLU) activation. In this learning strategy, there is a change in the learning goal, i.e., instead of learning the complete expected output, the residual is learnt. This makes more input values close to zero and more features zero after activation thus accelerating the learning process. The VDSR configuration is as shown in Fig. 2.

In this structure, the low resolution (LR) image undergoes bicubic interpolation to form the interpolated low resolution (ILR) image x. This input to the VDSR network goes through the convolutional and ReLU activation layers. VDSR predicts a residual

Table 1 Comparison of SRGAN and VDSR

DATASET	SRGAN		VDSR	
	PSNR (dB)	SSIM	PSNR (dB)	SSIM
SET 5	29.4	0.8472	37.53	0.9587
SET 14	28.49	0.8184	33.03	0.9124

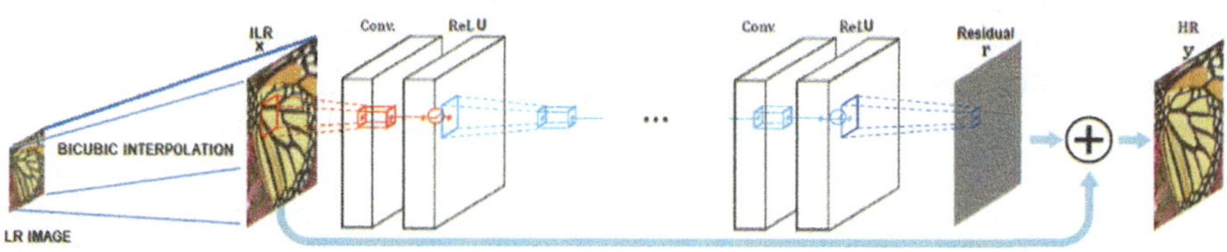

Fig. 2 VDSR network structure [19]

image r, and the addition of residual with interpolated low resolution image results in the desired high-resolution output y. This residual is given by

$$r = y - x \tag{1}$$

We have trained VDSR using luminance channel of periocular images because human vision is more receptive to changes in brightness (luminance) than to changes in color. Given the HR image y and ILR x, VDSR network is trained to predict the residual image (Eq. 1). This is done using the IAPRTC-12 Benchmark data [21] and MATLAB R2018b. Once the network learns to predict the residual image, HR images are obtained by the addition of predicted residual image with the upsampled LR image, i.e.,

$$Y_{\text{highres}} = Y_{\text{residual}} + Y_{\text{lowres}} \tag{2}$$

where Y_{highres}, Y_{residual}, and Y_{lowres} represent the luminance channels of the HR image y, residual image r and ILR image x, respectively.

3.3 CNN Feature Extraction

The periocular image from the VDSR network is fed to the CNN feature extraction block. The CNN used for feature extraction in our work is Alexnet [22]. It consists of five convolutional (conv.) layers and three fully connected (f_c) layers. In addition, there are ReLU and max-pooling layers.

Convolutional layer is the main building block of CNN. These layers comprise a series of kernels or filters that target at extracting local attributes from the input images. Each kernel generates a feature map which is produced by convolution operation. The first convolution layer acquires low-level features, and the succeeding convolutional layers extract higher-level features.

The nonlinear layers introduce nonlinearity into the CNN thus enabling the learning of complex models. In Alexnet, the nonlinear activation function used is ReLU. Given input x and output $f(x)$, it applies the following function:

$$f(x) = max(0, x) \tag{3}$$

The pooling or subsampling layer reduces the dimensionality of the feature maps. In Alexnet, max pooling is used. The max pooling returns the maximum output inside a rectangular neighborhood thus making the features robust to minute variations for formerly learned features.

The feature maps of the preceeding layer are convolved with learnable kernels at a convolution layer and passed to the activation function to form the output feature map. Each output combines convolutions with many input maps. This can be represented by equation (4). The CNN feature extraction of periocular images consists of a number of such steps, and each one is made of cascading the three aforementioned layers.

$$x_j{}^L = f\left(\sum_{i \in M_j} x_i{}^{L-1} * k_{ij}^L + b_j^L\right) \quad (4)$$

where $x_j{}^L$—outputs of Lth layer, k_{ij}^L—kernel for the Lth layer, b_j^L—additive biases for the Lth layer, and M_j—set of input maps

3.4 SVM Classification

The features extracted from Alexnet is then passed into the classification block. The periocular recognition task being a multiclass identification problem, we choose to perform classification using simple multiclass SVM [23], which is quite popular in image classification tasks because of its robustness, accuracy, and effectiveness while using small training set. Since SVM is mostly used for the binary classification problem, error-correcting output codes (ECOC)-based SVM is used for implementing the multiclass problem, whereby the direct multiclass task is split into several binary classification tasks, and the results from these indirect classifiers are then combined to give correct predictions of identity class.

4 Experimental Results

4.1 Database

For our experiments, we employ the UBIRIS v.2 database. The database developed by the University of Beira Interior, Portugal, contains 11,102 sRGB images of the eye regions from 261 subjects in the visible wavelength, at-a-distance (4–8 m) and on the move [6]. This database is of great interest researchers concerned with less constrained biometrics.

4.2 Performance Metric and Baseline Method

The rank-k identification indicates the number of correct identification that occurs in top-k matches [3]. This means that the correct class has to be in top-k probability for it to be counted as correct. The only case for a predicted class to be correct is if the latter has the highest output probability, i.e., when rank-1 accuracy is considered. This implies that rank-1 is the most precise measure of error because it ignores all guesses that are strictly incorrect, whereas a rank-k for $k > 1$ tolerates some error. Accordingly, we rely on rank-1 accuracy to report the performance of our work since it penalizes all guesses that are incorrect. Rank-1 accuracy is given by the ratio of number of correct classifications to the total number of predictions.

The baseline attribute descriptor we used for comparison is the phase-intensive local pattern (PILP) by Bakshi et al. [3]. This baseline work achieved rank-1 periocular recognition accuracy of 87.62%, the best reported in the literature with UBIRIS v.2.

4.3 Experimental Setup

The periocular images obtained from the UBIRIS v.2 database are grouped into 258 distinct classes with each class containing the periocular region of a particular individual. Thus, these 258 classes correspond to 258 distinct individuals. We select 70% of the data in each class randomly for training and the remaining 30% for testing.

In this work, we use Alexnet [22] to extract features from the periocular images. Since this network is trained using one million images of the ImageNet database, it is capable enough to extract the complex attributes of the periocular images. For feature extraction, we deploy the pretrained Alexnet model in MATLAB R2018b. We determine the rank-1 recognition accuracy for each layer, using the output of the Alexnet layers as attribute descriptors so as to explore the representation capability of the CNN layers for the periocular recognition task.

The multiclass SVM used in our work is equivalent to combining multiple binary SVM classifiers, and the number of class depends on the count of subjects in the periocular database.

4.4 Performance Analysis

With the output from each layer of CNN used as an attribute vector to characterize the periocular region, the performance of each layer is assessed using rank-1 recognition accuracy. We get maximum rank-1 accuracy of 91.47% at 'fc6' (first fully connected layer in Alexnet model in MATLAB R2018b). The difference in accuracy in the distinct layers of Alexnet owes to the distinct levels of features or activations learnt by them. The channels in earlier layers learn simple features like color, edges, etc., while those in deeper layers learn much more complex features like that of eyes.

Different levels of visual content are encoded by each layer of CNN [24]. Therefore, the appropriate way to view these is by independently plotting the contents as a two dimensional image. For visualization, we normalize the activations to gray-scale such that minimum activation is 0 and the maximum is 1. Such a mapping leads to white pixels representing strong positive activations and black representing the strong negative counterpart. In such a case, the gray pixels represent weak activations. The activations in the initial layers retain almost all of the information in the input image, while those in subsequent deeper layers contain less visual contents of the original image and more information pertaining to the class to which that particular image belongs. Fig. 3 shows activations in the layers 'conv1' (first convolutional layer), 'fc6,' and 'output.' From this, it is evident that for the initial layers (i.e., 'conv1'), the activation is visually very similar to that of the original image, which owes to the fact that they retain the majority of the input as such. The one corresponding to 'fc6' has strong white pixels around the center of the periocular region which is absent in the other cases, and this strong activation leads to high accuracy in the 'fc6' layer. Toward the final layers (eg. 'output'), the activation image is almost black indicating that these layers do not activate at all since there is nothing more to learn. Consequently, the recognition accuracy falls toward the final layers of the network as shown in Fig. 4.

We also compare the performance at each layer without super-resolution as well. Results are as shown in Fig. 4. Taking the above results into account, we see a significant boost of almost 30% in the recognition accuracy with super-resolution.

Comparison of the proposed method with the existing periocular recognition works using UBIRIS v.2 is shown in Table 2. From this, it is evident that our approach gives the most reliable recognition accuracy than the existing counterparts.

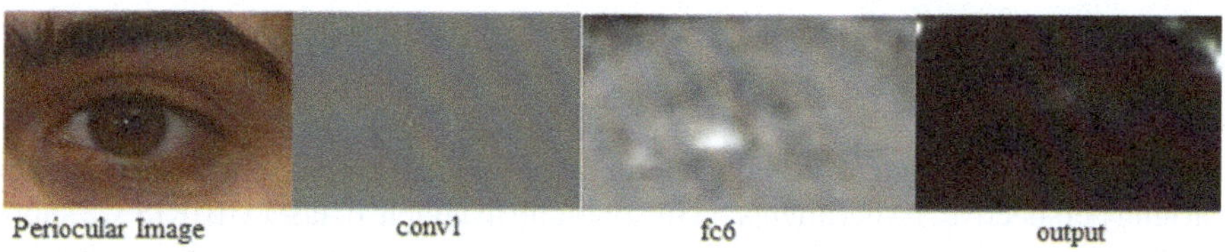

Fig. 3 Activations in Alexnet model

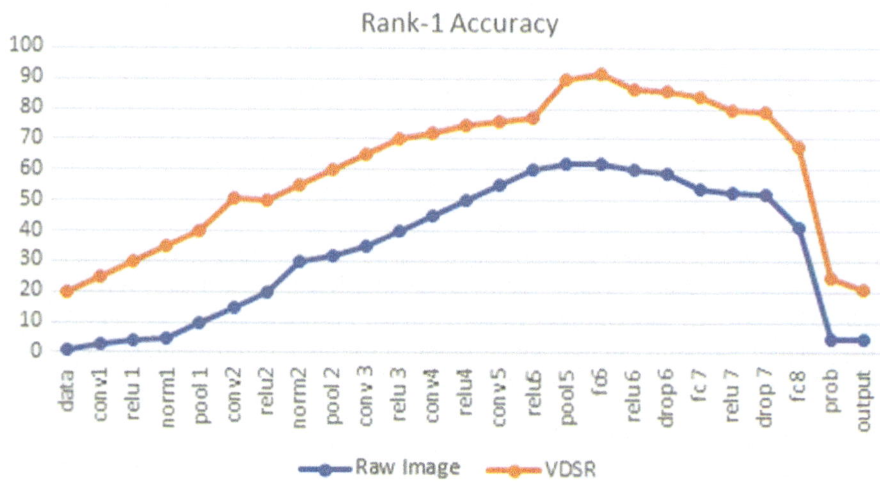

Fig. 4 Performance comparison

Table 2 Comparison with existing periocular recognition techniques

Works	Features	Rank-1 accuracy(%)
Bharadwaj et al. [25]	Circular Local Binary Patterns (CLBP), Global Descriptor (GIST)	73.65
Joshi et al. [26]	Local Binary Patterns (LBP)	81.03
Santos and Hoyle [27]	LBP, Scale-Invariant Feature Transform (SIFT)	56.4
Bakshi et al. [28]	Phase-Intensive Global Pattern (PIGP), CLBP, Walsh–Hadamard Transform-based LBP (WLBP)	82.86
Bakshi et al. [3]	Phase-Intensive Local Pattern (PILP), SIFT, Speeded Up Robust Features (SURF)	87.62
Proposed approach	Off-the-Shelf CNN features	91.47

5 Conclusion

In this work, we have addressed the approach of periocular recognition from a deep learning perspective. Experiments on the publicly available dataset UBIRIS v.2 show that the proposed work produces significantly better performance than the existing works. The proposed method enhances the recognition performance by the application of super-resolution which is more feasible than employing other advanced hardware solutions such as high-resolution image capturing devices and long focal lenses. This type of super-resolution could be extended to other biometric traits so as to improve the recognition or authentication accuracy. There are various open questions and challenges regarding the deployment of deep learning to the problem of periocular recognition, even though CNN such as Alexnet is effective in encoding discriminative features for periocular recognition. The computational intricacy of CNNs used for super-resolution as well as for recognition task can be addressed by using model reduction techniques such as pruning and compression [29]. There are several other frameworks in the field of deep learning such as recurrent neural network (RNN), stacked auto-encoder (SAE) [19], and so on. To improve the representation capability of periocular region, these architectures can be used independently or combined with conventional CNNs.

Acknowledgements This work is done as a part of the project Center of Excellence in Pattern and Image Analysis (CEPIA) 2019–20, which is funded by the Kerala State Planning Board.

References

1. Alonso-Fernandez F, Bigun J (2016) A survey on periocular biometrics research. Pattern Recog Lett 82:92–105
2. Park U, Reddy RJ, Jain AK, Ross A (2011) Periocular biometrics in the visible spectrum. IEEE Trans Inf Forensics Secur 6(1):96–106
3. Bakshi S, Sa PK, Majhi B (2015) A novel phase-intensive local pattern for periocular recognition under visible spectrum. Sci Dir Biocybernetics Biomed Eng 35(1):30–44
4. Nguyen K, Fookes C, Sridharan S, Tistarelli M, Nixon M (2018) Super-resolution for biometrics: a comprehensive survey. Sci Direct Pattern Recog 78:23–42
5. Vijayaraghavan K, Parpyani K, Thakwani SA, Aju D, Iyengar NChSN (2009) Methods of increasing spatial resolution of digital images with minimum detail loss and its applications. In: 5th international conference on image and graphics, Xi'an China, pp 685–689
6. Proenca H, Filipe S, Santos R, Oliveira J, Alexandre LA (2010) The UBIRIS.v2: a database of visible wavelength Iris images captured on-the-move and at-a-distance. IEEE Trans Pattern Anal Mach Intell 32(8):1529–1535
7. Hernandez-Diaz K, Alonso-Fernandez F, Bigun J (2018) Periocular recognition using CNN features off-the-shelf'. In: International conference of the Biometrics Special Interest Group
8. Ribeiro E, Uhl A, Alonso-Fernandez F (2018) Iris super-resolution using CNNs: is photorealism important to iris recognition? IET Biometrics. ISSN 2047–4938
9. Park U, Ross A, Jain AK (2009) Periocular biometrics in the visible spectrum: a feasibility study. In: Proceedings of the IEEE 3rd international conference on biometrics: theory, applications, and systems, pp 1–6
10. Lowe DG (2004) Distinctive image features from scale-invariant keypoints. Int J Comput Vis 60:91–110
11. Dalal N, Triggs B (2005) Histograms of oriented gradients for human detection. In: Proceedings of IEEE conference on computer vision and pattern recognition. San Diego, CA, pp 886–893
12. Ojala T, Pietikäinen M, Mäenpää T (2002) Multiresolution grayscale and rotation invariant texture classification with local binary patterns.' IEEE Trans Pattern Anal 971–987
13. Zhao Z, Kumar A (2017) Accurate periocular recognition under less constrained environment using semantics-assisted convolutional neural network. IEEE Trans Inf Forensic Secur 1017–1030
14. Proença H, Neves JC (2017) Deep-PRWIS: periocular recognition without the iris and sclera using deep learning frameworks. IEEE Trans Inf Forensic Secur 888–896
15. Tan C-W, Kumar A (2013) Towards online iris and periocular recognition under relaxed imaging constraints. IEEE Trans Image Process 22(10):3751–3765
16. Tapia J, Arellano C, Viedma I (2019) Sex-classification from cell-phones periocular Iris images. Selfie Biometrics Book (in Press)
17. Reddy N, Fahim Noor D, Li Z, Derakhshani R (2018) Multi-frame super resolution for ocular biometrics. IEEE/CVF conference on Computer Vision and Pattern Recognition Workshops (CVPRW)
18. Viola P, Jones MJ (2001) Rapid object detection using a boosted cascade of simple features, vol 1. In: Proceedings of the 2001 IEEE computer society conference on computer vision and pattern recognition, pp 511–518
19. Kim J, Lee JK, Lee KM (2016) Accurate image super-resolution using very deep convolutional networks. In: Proceedings of the IEEE conference on computer vision and pattern recognition, pp 1646–1654
20. Ledig C, Theis L, Huszár F, Caballero J, Cunningham A, Acosta A, Aitken A, Tejani A, Totz J, Wang Z, Shi W (2017) Photo-realistic single image super-resolution using a generative adversarial network. In: Proceedings of IEEE conference on computer vision and pattern recognition, Honolulu, USA, pp 105–114
21. Grubinger M, Clough P, Müller H, Deselaers T (2006) The IAPR TC-12 benchmark: a new evaluation resource for visual information systems, vol 5. In: Proceedings of the OntoImage 2006 language resources for content-based image retrieval. Genoa, Italy, p 10

22. Krizhevsky A, Sutskever I, Hinton GE (2012) Imagenet classification with deep convolutional neural networks. In: Pereira F, Burges CJC, Bottou L, Weinberger KQ (eds) Advances in neural information processing systems, vol 25. Curran Associates, Inc., pp 1097–1105
23. Scholkopf B, Smola AJ (2001) Learning with kernels: support vector machines, regularization, optimization, and beyond. MIT Press, USA
24. Nguyen K, Ross A (2017) Iris recognition with off-the-shelf CNN features: a deep learning perspective. IEEE Access
25. Bharadwaj S, Bhatt H, Vatsa M, Singh R (2010) Periocular biometrics: when iris recognition fails. In: 2010 fourth IEEE international conference on Biometrics: Theory, Applications and Systems (BTAS)
26. Joshi A, Gangwar A, Saquib Z (2012) Person recognition based on fusion of iris and periocular biometrics. In: 2012 12th international conference on Hybrid Intelligent Systems (HIS)
27. Santos G, Hoyle E (2012) A fusion approach to unconstrained iris recognition. Pattern Recogn Lett 33:984–990
28. Bakshi S, Pankaj KS, Majhi B (2014) Phase intensive global pattern for periocular recognition. In: 2014 annual IEEE India Conference (INDICON)
29. Cheng Y, Wang D, Zhou P, Zhang T (2019) A survey of model compression and acceleration for deep neural networks. IEEE Sig Process Mag Spec Issue Deep Learn Image Underst (Arxiv Extended Version)

Performance Evaluation of Recurrent Neural Networks for Short-Term Investment Decision in Stock Market

Alexandre P. da Silva, Silas S. L. Pereira, Mário W. L. Moreira, Joel J. P. C. Rodrigues, Ricardo A. L. Rabêlo, and Kashif Saleem

Abstract A significant challenge to invest in the stock market is to make complex decisions and evaluate different scenarios and trends in financial reports. An alternative capable of attending these needs is the use of algorithms for trading. This strategy may involve the use of artificial intelligence approaches to make effective decisions. Hence, this paper proposes the development of a solution to support decision making in the purchase and sale of stocks from the application of machine learning techniques. For this purpose, the performance of a recurrent neural network architecture, also known as long short-term memory (LSTM), was evaluated concerning its efficiency in the short-term prediction. The generated models were evaluated based on the root-mean-square error and mean absolute percentage error metrics. Results show the LSTM algorithm was capable of predicting stock prices with a low error approximately.

A. P. da Silva · S. S. L. Pereira · M. W. L. Moreira
Federal Institute of Education, Science, and Technology of Ceará, Aracati, CE, Brazil
e-mail: alexandre.bcc1@gmail.com

S. S. L. Pereira
e-mail: silas@lar.ifce.edu.br

M. W. L. Moreira
e-mail: mario.wedney@ifce.edu.br

J. J. P. C. Rodrigues (✉) · R. A. L. Rabêlo
Federal University of Piauí, Teresina, PI, Brazil
e-mail: joeljr@ieee.org

R. A. L. Rabêlo
e-mail: ricardoalr@ufpi.edu.br

J. J. P. C. Rodrigues
Instituto de Telecomunicações, Lisbon, Portugal

J. J. P. C. Rodrigues · K. Saleem
Center of Excellence in Information Assurance (CoEIA), King Saud University, Riyadh, Saudi Arabia
e-mail: ksaleem@ksu.edu.sa

S. M. Thampi et al. (eds.), *Applied Soft Computing and Communication Networks*, Lecture Notes in Networks and Systems 125,
https://doi.org/10.1007/978-981-15-3852-0_16

1 Introduction

The high volatility characterizes the financial markets, given the high level of uncertainties and risks associated with them [5]. Every day, stock exchanges begin their euphoric trading with constant price changes, which allows investors great opportunities. It is estimated that more than 3 trillion dollars will be transacted daily in the Forex market. With such a fickle and dangerous market, it is difficult for a layman or even a professional investor to track all price changes full time and, from analysis, to timely determine the best investment strategy for each occasion. In this way, the investment in the stock market brings with it the need to make complex decisions, to evaluate different scenarios and trends in the information contained in quotations in the stock exchanges, which requires, in addition to the specialist's expertise, a rigorous planning, as well as a risk management on invested capital.

Thus, it is prevalent in the investment world to distinguish four profiles of stock market strategies, to know, scalper, day trade, swing trading, and long-term investment. The scalper is characterized by speedy negotiations, which last on average between one and fifteen minutes. In this form of investment, investor accuracy, and speed in processing data related to the financial context of the trading company is of crucial importance. A day trade is, in turn, recognized for negotiations lasting up to one day. Like the scalper method, the day trade is an aggressive profile due to the high susceptible instability that a company's quote is subject to. Thus, both scalper and day trade are used by aggressive investors who expose themselves to more risk in their investments. In contrast to these investment strategy profiles, swing trade and long-term investments are characteristic of the conservative investment. This is because both swing trading, which deals with investments between three days to one month, and long-term investments, which deal with investments longer than one month, are essentially less risky to capital.

Given the individual characterization of an investor profile, the interpolation metric used for forecasting is an aspect that arises as a parameter for investment analysis. Currently, there are three well-known approaches for forecasting future price quotes in financial assets, to know, fundamental graphical, and technical analysis. As [4, 6], fundamental analysis is based on economic data as a determining factor for its prediction, so the related dataset to this analysis is closely linked to company balance and revenues, among other accounting information. This way, this study does not consider the use of stock price graphic resources in the decision-making process. In contrast to the fundamentalist approach, graphics analysis seeks to interpret the stock from data patterns and geometric shapes that, empirically, are identified in price quotation graphics of financial stocks. According to [4], this method is based on behavioral patterns that investors express operations on the stock market. These patterns are then represented in quotation charts by symbols. Following the concepts of fundamentalist and graphic analysis, the technical analysis aims to merge the foundations of both techniques. It admits that price behavior has a stochastic origin. Therefore, the probabilistic aspect is used as a tool to interpret the information. The use of statistical time series, as well as financial calculations, is of great importance in predicting stock price behavior. Liu et al. emphasize that the use of probability to support decision making in technical analysis puts this approach on another level, since it no longer fundamentals itself in the investor empiricism and its external factors as emotions, and is based only on statistical data [12].

The previously described strategies require a high level of study and expertise from user investors. This fact generates huge divergences among investors concerning the efficiency and accuracy of these techniques. Since these are experimental factors for analysis, there is no uniqueness in the obtained predictions. The majority of new investors have difficulties in understanding the dynamics of the financial stock market. A consequence is the distance of this investment modality, depriving these investors of monetizing their capital. Machine learning (ML) techniques have gained notoriety as candidates to supply this need, mainly with artificial neural networks (ANNs) and deep learning (DL) with recurrent neural networks (RNNs). According to [8], the application attractiveness of ANNs to conventional techniques is justified by the modeling capacity of more complex systems involving a large number of variables. ANNs are capable of solving problems of general-purpose in ML, such as approximation, classification, categorization, and prediction. In predicting time series, which is a task that involves the construction of models from observations in a given period, the application of classical models requires a deep understanding of the problem studied. In modeling with ANNs, since these are characterized by nonlinear and nonparametric modeling, the need to understand the process is reduced.

In parallel to the application of ML techniques in the economic environment, there are also several parametric programs, called expert advisors (EAs). These applications automatize strategies based on practical experience and perform stock operations based on rules previously established by the developer. A significant deficiency in EAs is the inability to detect changes occurring in the business perspective, which undermines their long-term effectiveness. Given the existing adaptability and prediction capabilities of ANNs, superior performance over EAs is expected, besides the consistency across the most diverse business environments.

Hence, this research aims at applying, analyzing, and assess ML techniques focused on ANNs to create a prediction model to support decision making in the purchase and sale of financial assets. The primary raw material for the development of the predictive models will be the commercial historical context, which is composed of information about openings, closures, maximums, and minimums of the financial assets investigated. The objective of this study is to evaluate a long short-term memory (LSTM) recurrent neural network model to anticipate strategic decisions related to the purchase and sale of financial assets from the historical context, in order to generate trading opportunities with low risk and a high potential for profit [7].

For this research, the present paper is organized as follows. The related works are reviewed in Sect. 2. Then, Sect. 3 describes the methodology used to develop the proposed solution. In Sect. 4, the results obtained are discussed. Finally, in Sect. 5, conclusions and future work are presented.

2 Related Work

In this section, a brief description of related studies found in the current state-of-the-art is presented. The use of ML as a tool for prediction has been explored extensively in recent years.

In [5], the authors presented an ensemble learning approach to predicting the next day closing price. It is based on a combination of three different models, *i.e.*, a recurrent neural network, known as LSTM, an autoregressive integrated moving average (ARIMA) model, and a sentiment analysis model. The combination purpose of the described techniques is to use the historical intuitions of the ARIMA and LSTM models. Besides, adding the current state of the market through the application of sentiment analysis. The output of each of these models is combined into a feed-forward ANN to predict the closing price of the stock. The authors argue that the proposed approach brings dynamism when considering technical and fundamental aspects of the financial market action.

In [20], the authors proposed a predictive model for trend forecasting of financial time series that is based on the combination of three ANNs. For a given financial asset, a network receives as input the dependent variables of this asset, which in the case study consists of opening, closure, maximum, minimum, and another network will use independent variables consisting of technical analysis indicators that derive from the treatment of dependent variables. The last network, called fusion ANN, is used to combine the output of the two previous models. The authors argue that this strategy aims to improve the prediction of the first two individually trained networks by aggregating knowledge. The experiments considered three actions of the Brazilian market (ITUB4, PETR4, and VALE5) to forecast uptrend or downtrend. In this process, the built architecture proposed only the forecast of the next business day trend, not exploring the interpolation of this interval. The authors used the results obtained by training the network and simulated with a test base possible results for the model application. A hit average of around 57% was obtained, considering the three assets analyzed. Regarding the financial return simulation, a balance of 13% was obtained, even in the current devaluation scenario in the test period. The authors conclude that the predictive system can generate higher profits at favorable times in the stock market.

Arvidsson et al. made a comparison of financial time series prediction capability from LSTM recurrent neural networks and traditional recurrent neural networks [2]. They investigate the ability of recurrent LSTM neural networks to model financial market behavior and its accuracy when compared to traditional models. For the execution of the experiments, this study used historical data of the asset *Euro–Dollar* (EUR/USD) in a graphical time of 10 min. Test data is composed of open, maximum, minimum, and close prices from February 9, 2017, to February 23 of the same year. The prediction modeling architecture using the LSTM network consisted of a 2048 cell input layer with a dropout of 20%. It is followed by the input layer, a hidden layer with 512 cells with a dropout of the same value as the previous one. A dense layer with linear type activation functions was added to the architecture, finishing the model. The models were configured to predict 10 min from a many-to-many LSTM architecture. The authors conclude that for a financial time series dataset, the LSTM model performs best in generalizing this problem category.

Complementing this idea, in [16], the authors present a comparison between traditional ML algorithms (regression, support vector machine (SVM), random forest, and ANNs) and an ensemble of LSTM networks with a traditional RNN in the task of predicting the financial time series. In this study, a set of recurrent LSTM network architectures are tested and evaluated for the stock market forecasting task. The results showed that the model based on RNNs had better performance when compared to conventional learning methods satisfying the previous studies presented. Furthermore, the authors argued that the generated RNN-LSTM-based model could be useful for both individual traders and corporate investors.

In [6], the authors propose a deep learning approach to predict stock prices in the financial market. The proposed approach involves predicting the price of an asset a week ahead, from a data sequence if a temporal series. The approach performed better when compared with state-of-the-art results. The proposed model combines the use of a recurrent neural network bidirectional LSTM and convolutional neural networks. CNN is used to get another sequence with a small size. The authors performed the experiments by the use of a dataset available on Yahoo Finances. They verify that the combination of convolutional layers and LSTM networks allows a better performance when compared with the SVM regressor.

The relevance of discussed research is notorious, given the state-of-the-art works related to the use of ML in predicting financial time series, as in [10, 12, 19, 21]. However, it is limited when extrapolated the inference of more extended periods. Few studies have investigated the behavior of long-term financial forecasts, and the literature related to the price forecasting domain in the stock market is significantly low. Thus, the present research seeks to differentiate from others by exploring the behavior of long-term forecasting and investigating the performance of ML techniques focusing on ANNs as a support agent for buying and selling financial assets on stock exchanges in the most diverse commercial panoramas.

3 Financial Market

Stock markets allow the purchase and sale of stocks issued by public and private companies. These companies provide a safe and efficient way for investors to be able to trade, either short- or long-term strategies, which ensures high liquidity daily. Exchanges contain a plethora of financial products such as government bonds, direct treasury, indexes, commodities, and many others. Among these products, the stock market, and currency exchanges are highly explored by investors. Forecasting of stock prices is a research field of interest by investors, banks, and entrepreneurs. For this task, mechanisms with distinct principles are used. The following prediction philosophies stand out, to know, fundamentalist, graphical, and technical analysis [4].

Fundamental analysis has economic foundations to infer predictions. The relationship among stocks, the financial health of companies and their financial panorama are the best data source to predict the future of stock prices and trends. These aspects encompass the field of microeconomics and macroeconomics. While microeconomics cares about regional balance aspects, the macroeconomics points to national and international questions. The primary resources for this analysis type are the balance sheets, production reports, expansion prospects, and administrative data such as salaries, payroll, and others [6]. Both graphical technical analyses are based on the ideas formulated by Charles Henry Dow, that were leveraged by Ralph Nelson Elliott's 1934 studies, published in his work "*The Wave Principle*." Graphical analysis shares the thought that the information necessary for the financial forecast of the stocks is denoted in the graphs that quote their prices. This mentality implies the expression of the behavior of economic agents through the recording of the movements of price quotations [4]. The technical analysis concerns a perspective based on statistical nature. Its management in forecasting is guided by probability theories and the use of correlation of financial variables. Price quotations of maximum and minimum values as well as indicators, which are the product of mathematical functions that attempt to measure the degree of volatility and financial volume, are used to model stock forecasts [12].

4 Long Short-Term Memory Neural Networks

ML is an area of artificial intelligence (AI) that aims to create computational learning techniques as well as the development of solutions that can learn automatically. Mitchell conceptualizes ML as a learning system that is capable of making decisions based on the experience gained [13]. Based on this, ML systems bring particular characteristics that allow the treatment of classification problems, prediction, classification, among others. A general classification of ML systems is related to their idea of their learning process, which can be supervised and unsupervised.

In terms of forecasting problems, the use of ANNs has gained much notoriety, especially with the evolution of computational performance that has occurred in recent decades. Thus, based on the idea of simulating the functioning of the nervous system of living beings in the problem-solving process, ANNs stand out for adapting to nonlinear problems. There are several ANNs architectures. Perceptron, Adaline degenerative, and recurrent are examples of structural arrangements used in problem-solving. Because of RNNs, these are actively used in regression tasks and problems in which sequence is one of the main factors for their resolution [9].

Deep learning is a subarea of ML that focuses on using ANNs to map intuitions between feature sets and categories [5]. Nowadays, deep neural networks are characterized by multiple layers and neurons, distinct and sophisticated ways of connecting layers and neurons, utilization of computational power available in current graphics processing boards for training, and automatic feature extraction. Thus, a deep neural network can be defined as a neural network with a large number of parameters and layers [15]. RNNs are ANNs that incorporate a feedback architecture, where their results are used for feeding. This means that the solutions obtained by their processing are current as they take into account previous values and not just independent answers as in traditional networks [9].

Forthcoming from Hopfield's recurrent neural network architecture, LSTM networks have similar characteristics in their operation. Its creation is due to the motivation to solve problems related to sequence, attribute of recurrent networks, but from a medium-term perspective. Unlike traditional recurring networks, which store short-term information, providing only the most recent context of data, LSTM networks seek to expand the problem foundation by setting the long-term situation [3]. This characteristic implies that in addition to the latest information stored on recurring networks, past information is also perpetuated in the processing cycle, giving it the long-term memory feature.

The specificity of the LSTM architecture is conferred by supporting the sequential capability of its structure, as shown in Fig. 1. Thus, it ensures the perpetuation of information for a dynamic period, which may be short term, when few cells are sequentially allocated, or long term, when many cells are sequentially disposed [17]. This operation is fostered by deep learning techniques, which theorize the parallel distribution in neural cell layers.

Figure 1 presents the recurrence of the information where, after processing the network, its output is redirected to feed it again. This process is perpetuated in cascade, and in addition to being fed back with the result of processing, it is passed on to the next layers.

In this way, the LSTM architecture model creates an information loop, stored in memory cells, which enables the expansion of contextual compliance to solve natural time series problems. This storage capacity enables the feedback of information at different levels. This causes the accumulation of knowledge, leading to the gain of information in its operation, especially in forecasting tasks.

The need for a long-term memory contextualization justifies the adoption of LSTM neural networks in the financial series prediction problem [11]. This solution is reiterated by the postulates of graphic analysis, where the pricing of stocks and financial securities are justified by their track record. Thus, given the short- and long-term knowledge capacity of the LSTM network, it presents itself as a potential candidate for the task of predicting financial behavior.

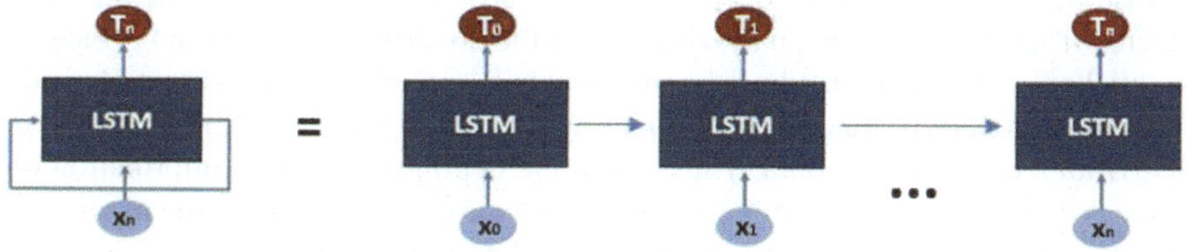

Fig. 1 LSTM network

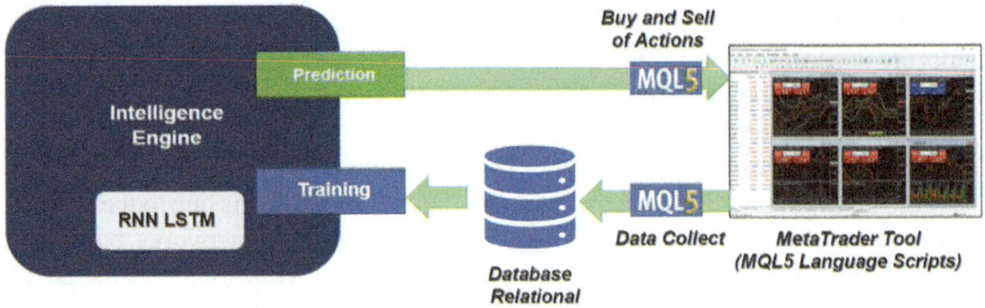

Fig. 2 Proposed solution architecture

5 Methodology

This is ongoing research that is characterized by the analysis of financial assets (equities) and the application of ML models to forecast trends for useful insights. The data used for analysis and forecasting is in the public domain, such as stock price quotes of different companies on the stock exchange.

5.1 *Proposed Solution Architecture*

The proposed prediction system architecture diagram is described in Fig. 2. The MetaTrader[1] tool was initially used for automating asset capture and executing buy and sell orders, enabling automated scripting using the MQL4 language. The collected data were stored in a relational database, which was used to build the suggested predictive model, integrating an intelligence engine in the form of an application programming interface (API).

5.2 *Data Preparation*

The data preprocessing phase aims to build the final dataset, which a priori consists of the historical quotation of the openings of the Euro–Dollar asset. This attribute was selected from the graphical analysis advocated ideas that give the opening price an essential piece for understanding the financial behavior of an asset. This study used a script developed for this purpose, which will be executed in the trading platform *MetaTrader* to collect the data. Thus, given the continuous flow of information on business days, obtaining real-time asset opening quote prices enables the prediction process to be automated [1]. From these raw data initially collected and required

[1] https://www.metatrader5.com/.

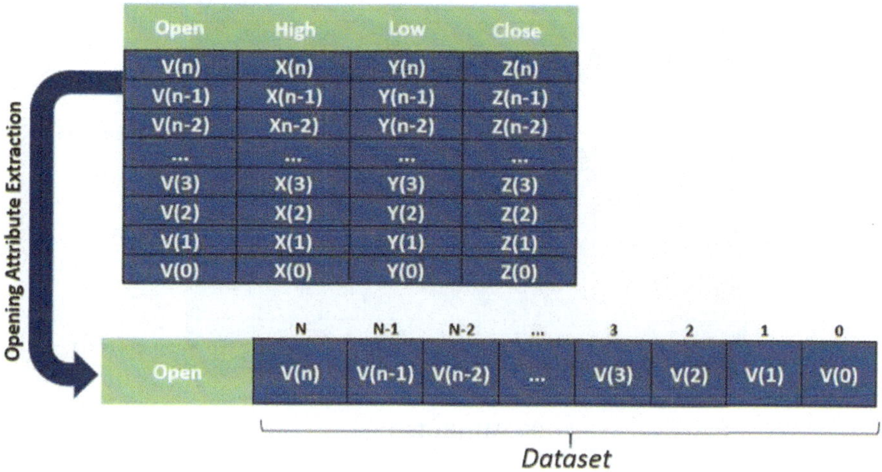

Fig. 3 Financial dataset extracted from MetaTrader software

for the predictions, the opening attribute extraction is done as represented by Fig. 3. Then, the process of data integrity verification begins. At this stage, data quality, outliers, and data processing, such as discretization and normalization, are checked.

After the data processing step, each attribute of the resulting dataset, indicated by the Fig. 3, must be scaled using the Min-Max normalization, available in the scikit-learn library [18], which is expressed by Formula (1).

$$x_{\text{scaled}} = \frac{x - x_{\min}}{x_{\max} - x_{\min}} \tag{1}$$

In [12, 14], the authors present prediction models that use LSTM network architectures. In these studies, the prediction is made to infer one future element at a time. However, given the need to estimate more than one element, this research evaluates an LSTM-based architecture that predicts more than one period. Therefore, considering the temporal integrity of the sample, the data contained in the dataset were manipulated.

Because a time series has indices ranging from 1 to t, which represent the first and last elements collected in ascending order respectively, where the first element collected is the oldest, the forecast goal is the set of elements whose rates range from $t + 1$ to $t + n$. Therefore, based on the dataset, this study aims to predict the elements $t + 1$, $t + 2$, and $t + 3$. For this reason, it was considered to change the periodicity of the time series to obtain the objective forecasts. In this process, starting from the dataset, three datasets were built, to know, A, B, and C. The dataset A consists of the dataset replica. This is due to the time unit of the data; therefore, it is possible to estimate the $t + 1$ element as shown in Fig. 4. The same procedure is performed to create datasets B and C. In these, the granularity was adjusted to match the target

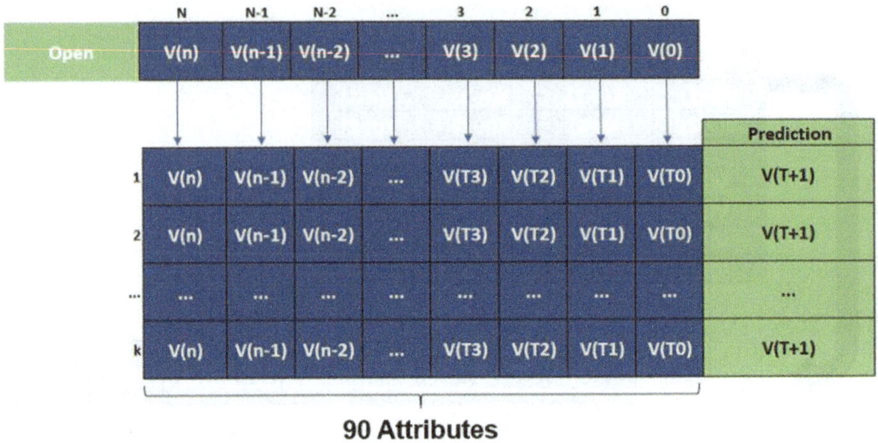

Figure 4 Training dataset for long-term memory network A

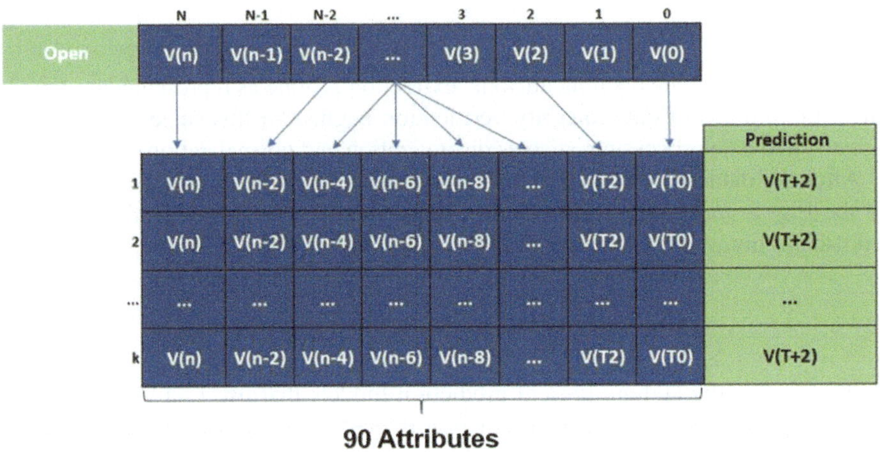

Figure 5 Training dataset for long-term memory network B

forecast. Therefore, the dataset was partitioned at regular intervals of 2 and 3 periods. Figure 5 shows this procedure for dataset B.

From the three datasets, three LSTM neural network structures were modeled, which will allow the prediction of three future elements, namely the $t + 1$, $t + 2$, and $t + 3$ elements.

Table 1 Neural network architecture

Layer	Type	Output shape	Params
Input layer	(LSTM)	(90, 100)	40800
Hidden layer 1	(LSTM)	(90, 50)	30200
Dropout 1 (Dropout)	(90, 100)	0	
Hidden layer 2	(LSTM)	(90, 50)	20200
Dropout 2 (Dropout)	(90, 50)	0	
Hidden layer 3	(LSTM)	(50)	20200
Dropout 3 (Dropout)	(50)	0	
Output layer	(Dense)	(1)	51

5.3 Experiments

For the evaluation of ML techniques, external libraries, such as scikit-learn [18] and the Python programming language for data science tasks, were used, allowing rapid prototyping in mining tasks. The Keras[2] library was used to allow the construction of ANNs at a high level, enabling the processing via TensorFlow.[3]

The root-mean-square error (RMSE) and mean absolute percentage error (MAPE), described in Formulas (1) and (2), were used to analyze the performance of the models.

$$\text{RMSE} = \sqrt{\frac{1}{n}\sum_{i=1}^{n}\left(\frac{d_i - f_i}{\sigma_i}\right)^2} \tag{2}$$

$$\text{MAPE} = \frac{1}{n}\sum_{t=1}^{n}\left|\frac{A_t - F_t}{A_t}\right| * 100 \tag{3}$$

The ANN architecture used for this study consists of an input layer which receives an input vector with 90 attributes, referring to each instance in the datasets. These attributes describe the last 90 price open values of the analyzed stock. The network architecture is composed of three hidden layers, each one composed of 50 nodes and an output layer with only one node, which will provide the estimated value for a stock. The activation function used was linear, described by the expression $f(x) = ax$. This study also applied a dropout of 30% in each hidden layer to avoid overfitting. Table 1 presents the neural network architecture. This network structure was chosen arbitrary and contains a total of 111,451 trainable parameters.

[2]https://keras.io.

[3]http://tensorflow.org/.

6 Results and Discussion

From the performance of the training and evaluation stages of predictive models, this section presents the individual results of the predictions obtained after normalization of the data, and finally, the combined result of the predictions. The tests performed to evaluate the accuracy of the model are based on the data of the Euro–Dollar assets, which run from September 2018 to July 2019 (Fig. 6).

Individually, it is possible to perceive that the results obtained by the forecasts generalized well to the current scenario. Network A obtained the highest absolute average error of 0.0298, while network B and C obtained errors of 0.0198 and 0.0195, as presented in Figs. 7, 8 and 9.

The RMSE and MAPE performance metrics are represented in Table 2. The neural network A presented many noises. Although the model adapted well, when compared with the B and C networks, its results were lower than expected. Due to this, the model predictions were substantially lower than expected, even though the RMSE and MAPE were within the acceptable range as shown in Table 2.

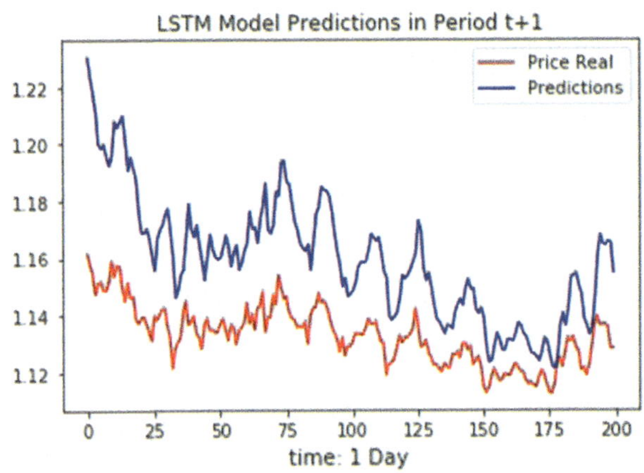

Figure 6 Set of network A forecasts

Table 2 Evaluation of long short-term memory networks

Metric	Model A	Model B	Model C
RMSE	0.000861	0.000006	0.0000055
MAPE	2.319	0.54187	0.5179

Figure 7 Set of network B forecasts

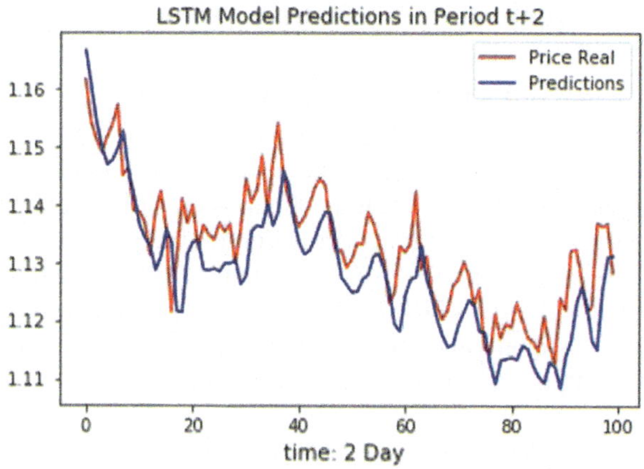

Figure 8 Set of network C forecasts

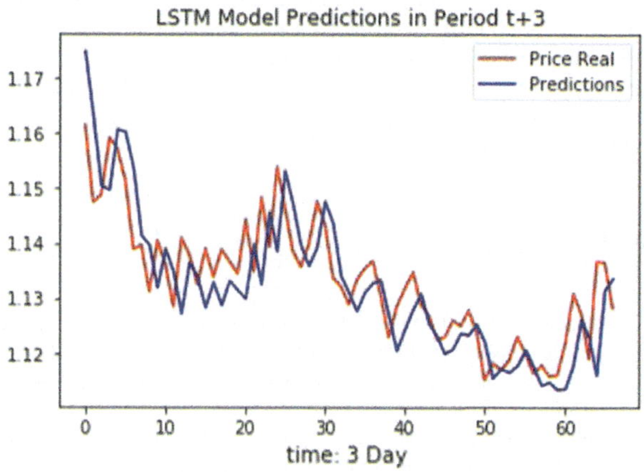

Figure 9 Model forecast set

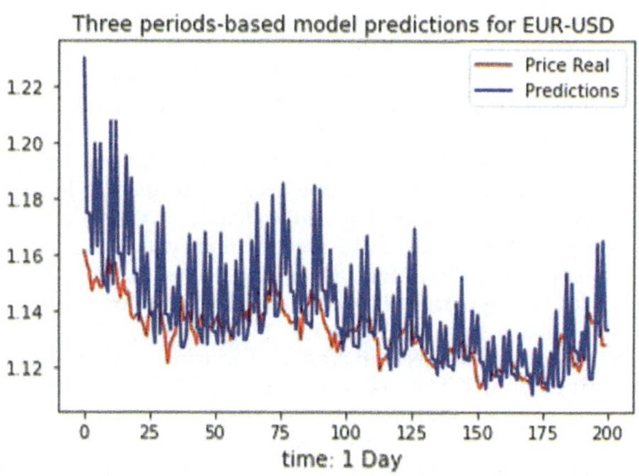

7 Conclusion and Future Work

The use of LSTM neural networks in the process of forecasting financial time series has demonstrated great potential for inference in the short term. The proposed architecture application to infer more than one element is therefore still the object of analysis since the results obtained are still not satisfactory to effectively aid in a negotiation, although it clearly defines the analyzed asset trend.

Further works involve searching for the combination of the models described with other techniques of ML, in order to obtain better predictions. Besides, it is planned to investigate other strategies on the behavior of forecasting financial series in the short term with multiple periods ahead.

Acknowledgements This work has been supported by the Brazilian National Council for Scientific and Technological Development (CNPq) via Grants No. 201155/2015-0 and 309335/2017-5; by the National Funding from the FCT—Fundação para a Ciência e Tecnologia through the UID/EEA/500008/2019 Project; by the International Scientific Partnership Program ISPP at King Saud University through ISPP 0129; and by the Federal Institute of Education, Science and Technology of Ceará (IFCE), via the institutional program of scientific initiation scholarships (PIBIC), Edital No. 1/2019.

References

1. Aidemark J, Askenäs L (2018) The creation of users: a learning experience in information system development. In J E-Health Med Commun 9(2):74–88
2. Arvidsson P, Ånhed T (2017) Sequence-to-sequence learning of financial time series in algorithmic trading
3. Boccato L (2013) Novas propostas e aplicações de redes neurais com estados de eco. Ph.D. thesis, Universidade Estadual de Campinas, Unicamp, SP
4. Cervelló-Royo R, Guijarro F, Michniuk K (2015) Stock market trading rule based on pattern recognition and technical analysis: forecasting the DJIA index with intraday data. Expert Syst Appl 42(14):5963–5975
5. Deorukhkar OS, Lokhande SH, Nayak VR, Chougule AA (2019) Stock price prediction using combination of LSTM neural networks, ARIMA and sentiment analysis. Int Res J Eng Technol 6(3):3497–3503
6. Eapen J, Bein D, Verma A (2019) Novel deep learning model with CNN and bi-directional LSTM for improved stock market index prediction. In: 2019 IEEE 9th annual computing and communication workshop and conference (CCWC). IEEE, pp 0264–0270
7. Fischer T, Krauss C (2018) Deep learning with long short-term memory networks for financial market predictions. Eur J Oper Res 270(2):654–669
8. Hatcher WG, Yu W (2018) A survey of deep learning: platforms, applications and emerging research trends. IEEE Access 6:24411–24432
9. Haykin S (2001) Neural networks: principles and practice. Bookman
10. Hiransha M, Ab Gopalakrishnan E, Menon VK, Soman KP (2018) NSE stock market prediction using deep-learning models. Procedia Comput Sci 132:1351–1362
11. Kim HY, Won CH (2018) Forecasting the volatility of stock price index: a hybrid model integrating LSTM with multiple GARCH-type models. Expert Syst Appl 103:25–37
12. Liu S, Liao G, Ding Y (2018) Stock transaction prediction modeling and analysis based on LSTM. In: 13th IEEE conference on industrial electronics and applications (ICIEA), May 31–June 2, Wuhan, China. IEEE, pp 2787–2790

13. Mitchell TM (1997) Machine learning. McGraw Hill, Burr Ridge, IL
14. Pang X, Zhou Y, Wang P et al (2018) An innovative neural network approach for stock market prediction. J Supercomput (in press):1–21
15. Patterson J, Gibson A (2017) Deep learning: a practitioner's approach. O'Reilly Media, Inc
16. Pawar K, Jalem RS, Tiwari V (2018) Stock market price prediction using LSTM RNN. Springer, Singapore, pp 493–503
17. Pawar K, Jalem RS, Tiwari V (2019) Stock market price prediction using LSTM RNN. In: Emerging trends in expert applications and security. Springer, pp 493–503
18. Pedregosa F, Varoquaux G, Gramfort A et al (2011) Scikit-learn: machine learning in python. J Mach Learn Res 12:2825–2830
19. Sai Reddy VK (2018) Stock market prediction using machine learning. Int Res J Eng Technol (IRJET) 5(10):1033–1035
20. Thiele CC, Adami AG (2016) Previsão de séries temporais financeiras: modelo baseado em redes neurais artificiais. Revista Brasileira de Computação Aplicada 8(2):113–128
21. Zhang K, Zhong G, Dong J, Wang S, Wang Yong (2019) Stock market prediction based on generative adversarial network. Procedia Comput Sci 147:400–406

The manufacturer's authorised representative in the EU is Springer Nature Customer Service Centre GmbH, Europaplatz 3, 69115 Heidelberg, Germany. If you have any concerns regarding our products, please contact ProductSafety@springernature.com

Printed and bound by CPI Group (UK) Ltd, Croydon, CR0 4YY
23/07/2026
02174995-0004